T0177797

SELFISH GENES TO
SOCIAL BEINGS

SELFISH GENES TO SOCIAL BEINGS

A COOPERATIVE HISTORY OF LIFE

JONATHAN SILVERTOWN

OXFORD

UNIVERSITY PRESS

OXFORD
UNIVERSITY PRESS

Great Clarendon Street, Oxford, OX2 6DP,
United Kingdom

Oxford University Press is a department of the University of Oxford.
It furthers the University's objective of excellence in research, scholarship,
and education by publishing worldwide. Oxford is a registered trade mark of
Oxford University Press in the UK and in certain other countries

Published in the United States of America by Oxford University Press
198 Madison Avenue, New York, NY 10016, United States of America

British Library Cataloguing in Publication Data
Data available

Library of Congress Control Number: 2023947247

ISBN 978–0–19–887639–7

DOI: 10.1093/oso/9780198876397.001.0001

Printed and bound in the UK by
Clays Ltd, Elcograf S.p.A.

Links to third party websites are provided by Oxford in good faith and
for information only. Oxford disclaims any responsibility for the materials
contained in any third party website referenced in this work.

To my colleagues in the University and College Union in their collective struggle for fair treatment.

Contents

IV. GENES

List of Illustrations

List of Plates

PART I

Groups

I

Topos and *Narcos*

On 19 September 1985 a magnitude eight earthquake struck Mexico City, the strongest quake yet recorded. In minutes, thousands of build-ings collapsed or were severely damaged including a hospital tower of 12 stories where more than 500 people died.[1] Mexico City is built on the bed of a lake, now mostly drained but affording protection to the city in Aztec times. As the shock wave hit, the dried-out lake sediments behaved like the mattress of a waterbed; undulating, toppling some buildings, and leaving others right next to them standing. Some taller buildings remained upright but pancaked with upper or lower stories collapsing while other floors held up. Whether you lived or died depended on which floor you happened to be on at the time.

Death is a common motif in Mexican folk culture, often depicted as a grinning skeleton in a sombrero playing a guitar. That 19 September, death's fingers plucked many thousands in a heartbeat. Ten thousand may have died, though the true death toll is not known. Times of crisis reveal the best and the worst in both people and institutions. The Mexican authorities were paralysed, but the people of the city took the rescue of their neighbours into their own hands. In the critical few days immediately after the quake, citizens rescued 4,000 people from beneath the rubble. In the first four hours, volunteers moved nearly 2,000 critically ill patients from damaged hospitals to safety.

Among the groups of citizens who came to the rescue of those buried alive in the rubble of Mexico City was a group of youths from the district of Tlatelolco, a tough neighbourhood in a tough city. It was a point of pride for them to crawl into buildings where no one else dared to go. They had no training and no specialized equipment, but in a crisis, they found the will and the means to do what had to be done. A few months after the quake, the group formed a permanent team of volunteers to train for and provide

Figure 1 The *Topos de Tlatelolco*

rescue in civil emergencies like the 1985 earthquake. The *Topos de Tlatelolco* (Tlatelolco moles) (Figure 1) have since helped rescue people in disasters throughout Latin America and across the globe from Italy to Indonesia. Their list of requirements for prospective recruits reads like a definition of altruism. Aspiring *Topos* must want to help others, be prepared to train hard, go to the rescue of anyone who needs it, and not receive personal gain or expect any form of recognition.[2]

Who would deny that the *Topos* and the whole community response to the 1985 Mexico City earthquake represents anything less than the very best in human nature? And yet today it is an understatement to say that Mexico is a troubled country. *Narcotráficos* kill at will and perpetrate untrammelled violence. On a single day in June 2020 there were 117 murders, setting a new daily record.[3] Femicide is increasing, with several very gruesome killings in 2020 and escalating rates of violence against women.[4] And yet Mexico is not unique. In 2020 it stood only fourth among countries ranked by murder rate.[5]

For centuries, philosophers have argued about whether it is human nature to be kind and good, or to be selfishly bad. Is untamed society a war of each against all that must be policed by a powerful ruler, as Thomas Hobbes

(1588–1679) believed? Or, in our natural state, are we noble savages, when not spoiled by civilization, as imagined by Jean-Jacques Rousseau (1712–1778)? So, which is the real us? In our deepest natures are we *Topos* or *Narcos*? Or is this perennial question the wrong one to ask ourselves since each of us is capable in some degree of both kindness and cruelty? In the words of Alexander Solzhenitsyn in *The Gulag Archipelago*:

> If only there were evil people somewhere insidiously committing evil deeds, and it were necessary only to separate them from the rest of us and destroy them. But the line dividing good and evil cuts through the heart of every human being.

The example of Mexico and so many others like it show that human nature is neither essentially good nor essentially bad, but generally cooperative. Sometimes, perhaps most of the time, we cooperate in ways that are peaceful and good or, like the *Topos*, even heroic, but if we cooperate as the *narcotráficos* do in organized crime, the results are monstrous. Hobbes wanted us to believe that we are naturally evil. For Rousseau, on the other hand, we are the reverse: innocent beings corrupted by enlightenment. Neither extreme view of human nature can be correct if you look at how we actually behave. Cooperation is in our nature, for good *and* ill.[6]

'A cooperative species', 'super-cooperators', 'an ultrasocial species'— writers on this subject outbid each other in trying to describe just how cooperative we are, and there is little doubt that superlatives are justified.[7] Think of how much cooperation had to occur before you could read this book: the invention of language and then writing; the development and practice of science; the dissemination of knowledge via books and the internet; your own education and mine. Underlying all of this is the division of labour and an economy that efficiently supplies us with food, sparing us time and energy for more than the mere exigencies of life. Truly, we live in a thicket of teams perched on a mountain of cooperation.

This self-knowledge is liberating. We need no longer fret that human nature is sinful or fear that the milk of human kindness will run dry. Yet there is still a nagging doubt that something biological in us compels us to be selfish: our genes. Richard Dawkins' landmark book *The Selfish Gene* contained the plea: 'Let us try to teach generosity and altruism because we are born selfish.' He later retracted this statement and admonished that the book itself, for those who read past the memorable title, actually describes how genes cooperate in their own interests.[8] That, as we shall see, is a recurring pattern. Genes are selfish in the sense that they are driven by self-replication, and yet cooperation continually rears its head. Natural selection is a race into the future where individual success is measured in the number of descendants. How is it possible that in such a race, natural selection can be a team sport?

Humans are fundamentally team players, but could it be that all of nature has been teaming up since the dawn of life four billion years ago?[9] Teams are groups of cooperators, invariably bound together by the benefits that accrue from force of numbers and a division of labour. All teams, whether they play football or make footballers, have a set of properties in common. The parallel between a football team and a team of cells is not merely metaphorical. One can imagine a team of individuals that all behave identically, but actual teams are never like that. Team members play different roles that have different functions. The goalie and the white blood cell both have specialized defensive functions in their teams. The goalie defends the goal, the white blood cell defends the goalie.

Team formation requires holding rivalries and self-interest in check. The division of labour in a team requires management for success. A major responsibility of the football manager is to constrain the ambitions of individual players for the overriding benefit of the team's success. Each cell in the body contains a copy of the individual's genome and this serves the same managerial function. Occasionally there are mutations and cells escape control and begin to multiply—creating tumours. The most dangerous tumours are those that recruit their own blood supply, accumulate multiple cell types that team up, and enable the cancer to spread around the body. Even rogue cells benefit from teamwork.

So, perhaps the first team players were not apes, bees, or even bacteria. Could the first teams have been the naked molecules that were the ancestors of genes at the very origin of life? Surely genes, though notoriously selfish in their impetus to propagate more of their own kind, must team up too. If they didn't, wouldn't life still be a primordial alphabet soup? And ultimately, is the history of life in fact a story of cooperation? Stick with me to find out.

We start at the large scale and with the familiar: the behaviour of groups in our own species. Through the four major parts of the book, we descend in scale and go back in time as we shift focus from groups to individuals, to cells, and then to genes. We investigate whether the rules of cooperation that we encounter in our daily lives are fundamentally the same as those that apply to other species; to how our cells cooperate within the body, how the parts of the cell work together, and how selfish genes cooperate to make social beings. The wanton violence that our species is so prone to is the hardest test, so I start in the unlikeliest of places for a history of cooperation: on the Western Front during the First World War.

The brutality and carnage are keenly remembered, even a century later when all living witnesses have died. Between 1914 and 1918, the First World War robbed even the smallest villages in the combatant nations and their colonies of sons, fathers, and husbands in a war that was supposed to end all wars. Countless women and children also died, unrecorded on war memorials. The bloodbath of the war killed at least 8.5 million combatants and perhaps 13 million civilians.[10] The scale of the killing was industrial and so indiscriminate that only rough estimates are possible. In the battle of the Somme, the British army alone lost more than 57,000 soldiers in a single day. For comparison, this is nearly the same as the number of American soldiers who were killed in the 14 years the USA fought the war in Vietnam. Such huge losses resulted from futile and repeated attempts to break the stalemate that existed along the Western Front, where opposing forces were dug into trenches, confronting each other along a line stretching through northern France and Belgium.

If we are to learn anything from the senseless slaughter, it should be that even in such dire circumstances, cooperation can spontaneously emerge between enemies. British private Marmaduke Walkinton recalled years afterwards what happened on Christmas Eve, 2014:

> We were in the front line; we were about 300 yards from the Germans. And we had, I think on Christmas Eve, we'd been singing carols and this that and the other, and the Germans had been doing the same. And we'd been shouting to each other, sometimes rude remarks more often just joking remarks. Anyway, eventually a German said, 'Tomorrow you no shoot, we no shoot'. And the morning came and we didn't shoot and they didn't shoot. So then we began to pop our heads over the side and jump down quickly in case they shot but they didn't shoot. And then we saw a German standing up, waving his arms and we didn't shoot and so on, and so it gradually grew.[11]

In many places along the Western Front, soldiers from the opposing sides agreed an informal truce that continued for two days. Enemies fraternized in the no man's land separating their trenches, swapping gifts and mementoes (Figure 2). Football matches were played. A German artillery officer named Rickner described meeting French soldiers:

Figure 2 The Christmas Truce of 1914

> I remember very well Christmas, I remember the Christmas Day when the German and the French soldiers left their trenches, went to the barbed wire between them with champagne and cigarettes in their hands and had feelings of fraternisation and shouted they wanted to finish the war.

If it had been up to the soldiers in the trenches, the Christmas truce would have become a lasting peace, but word soon got back to the generals behind the lines and on both sides, they commanded it end.

> The generals behind must've seen it and got a bit suspicious so what they did, they gave orders for a battery of guns behind us to fire, and a machine gun to open out and officers to fire their revolvers at the Jerries. 'Course that started the war again. Ooh we were cursing them to hell, cursing the generals and that, you want to get up here in this stuff never mind your giving orders, in your big chateaux and driving about in your big cars. We hated the sight of the bloody generals. George Ashurst

The now famous Christmas truce of 1914 was not to be repeated during the rest of the war, but the conditions that generated it persisted. Soldiers on either side, never separated by more than shouting distance, knew that their adversaries were just like them, suffered the same privations, and risked a futile death just as they did. These foundations of empathy and understanding led to a system of live-and-let-live along the Western Front, wherever the high command's attention was focused elsewhere.[12]

In these quieter places, soldiers on both sides realized that they and their adversaries still needed to put on a show of aggression for the benefit of their superior officers. On the German side, the Saxon regiments were especially cooperative and, by means of a message tied to a stone lobbed into the trench, they informed the 46th division:

> We are going to send a 40 lb bomb. We have got to do this, but don't want to.
> It will come this evening, and we will whistle first to warn you.

A newly arriving battalion of the 51st division heard from the trench opposite:

> We Saxons, you Anglo Saxons, don't shoot!

A bit later, when the same Saxon unit was about to be relieved by Prussians, they shouted to the English, 'Give them hell!'

The artillery on both sides were ranged at a greater distance from the front lines and continually threatened the cooperation that developed between the infantry in the trenches who were separated from the enemy by only by the width of no man's land. A British officer recalled this incident when facing a Saxon unit:

> I was having tea with A Company when we heard a lot of shouting and went out to investigate. We found our men and the Germans standing on their respective parapets. Suddenly a salvo arrived but did no damage. Naturally, both sides got down and our men started swearing at the Germans, when all at once a brave German got on to his parapet and shouted out 'We are very sorry about that; we hope no one was hurt. It is not our fault, it is that damned Prussian artillery.'[13]

Messages were sometimes exchanged on signs raised above the trench, such as one that read: 'Don't fire, East Surreys, you shoot too well.' Defused armaments were also used to deliver messages but must have taken a good deal of trust to read. The poet Robert Graves reported that corporals in his battalion of Royal Welch Fusiliers received an invitation to dinner from their German counterparts inside a defused grenade.

Sign boards and friendly messages tied to missiles were all very well but rather obvious. The appearance of aggression had to be maintained for the benefit of the officers, even if some of them turned a blind eye. So, soldiers ritualized their fire against the enemy's positions, knowing that their cooperative intentions would be understood from the predictability of their attacks which rendered them harmless. A Colonel Jones, who served in a unit of the 48th division, discovered within a couple of days of arriving at the front that the Germans shelling his trench followed a fixed routine each day, so that he could predict exactly where and when the next shell would hit. This enabled him to show off in front of senior officers by taking risks that looked very dangerous to them, but which he knew were quite safe. If he didn't deserve a medal for bravery, he certainly deserved one for the trust he placed in the enemy's intentions.

When soldiers in the front line were relieved by comrades from the rear, they passed on the local rules to them, such as:

> [The French allies] explained to me that they had practically a code which the enemy well understood: they fired two shots for each one that came over, but never fired first.[14]

These details of the friendlier side of trench warfare on the Western Front come from a remarkable study by historian Tony Ashworth in which he describes the live-and-let-live system based on first-hand accounts by soldiers on all sides of the conflict.[15] Ashworth's study is of much more than historical interest because it is a paradigm case of how cooperation can emerge between antagonists, following conditions that apply in a wide range of other situations from the evolution of animal conflict to market economics.

To understand the conditions that can lead to cooperation between adversaries, we can generalize what occurred on the Western Front as a contest between opponents, each of whom has to decide whether or not to cooperate with the other side. In formal terms this is a game because, like in chess, how one player should act to win is determined by how the other player behaves. This formal use of the term 'game' does not imply a sport or that the outcome is trivial. This particular type of game, known as the prisoner's dilemma, differs from games such as chess because cooperation makes it possible for both players to win. By contrast, in a game such as chess for one player to win the other must lose. The sum of a win ($+1$) and a loss (-1) is zero, so chess is called a 'zero-sum game'.

The stakes on the Western Front were for life or death and the survival of both opponents was clearly a win-win. Another difference from chess is that players in the prisoner's dilemma decide in each round what their action will be and then they both make their move simultaneously. On the Western Front, cooperation meant not attacking, or if firing, then doing so to avoid casualties. There are four possible combinations of actions (scenarios) for two players of this game—Western Front style (Table 1).

Table I. The four possible combinations of play, or scenarios, in a round of prisoner's dilemma—Western Front style.

	Action
Britain and allies	**Central Powers including Germany**
Cooperate	Cooperate
Cooperate	Attack
Attack	Cooperate
Attack	Attack

Each action has a different consequence, or payoff, depending upon what the opponent does, and we can rank these from the best outcome to the worst for any given player, Brit or German. To start, note that attacking your opponent when they cooperate must bring the highest payoff because if you can kill your enemy before they kill you, you have won. This payoff is called T, for the Temptation to attack (i.e. not cooperate). Conversely, being on the receiving end of an attack when you yourself decided to cooperate must be the worst result because you wind up dead. This is called the Sucker's payoff (S).

So, we have established the highest (T) and lowest (S) payoffs among the four scenarios, but which of the two remaining ones occupy second and third place? Mutual cooperation should logically be in second place because neither side receives damage from the other. This payoff is called the Reward for cooperation and according to the rules of the prisoner's dilemma, R must be worth *more* than the average of T + S. This condition is a way of saying that this is not a zero-sum game (if I win you must lose), because mutual cooperation produces a win-win. To complete the ranking of scenarios, the consequences of mutual attack must logically be worse than for mutual cooperation because in the mutual attack each opponent damages the other. The third-placed payoff is therefore the Penalty (P) for mutual attack.

In the basic version of this game, it is played as a single, stand-alone round. When that happens, you have no information as to what your opponent might do and the safest assumption about the guy pointing a machine gun is that he's going to kill you if you don't kill him first. He, of course, is thinking the same about you. The predicted result is therefore mutual attack and mutual injury. If you are contemplating your strategy beforehand, you are on the horns of a dilemma. The rational move is to attack because this is what you can expect your opponent to do and you can't risk being suckered by cooperating. However, the dilemma is that mutual cooperation has a higher payoff than mutual attack (R > P). Dare you take the risk?

Attack rather than cooperation was certainly in the minds of the generals who planned operations on the Western Front at the start of the war. Both sides were convinced they would win and that it would all be over by Christmas 1914. Big mistake. Once the two sides were dug into trenches, the situation changed in two very important ways. First, because there was a stalemate, confrontation would be repeated, and no one knew when it would end. Second, there was communication across no man's land between the infantry on each side, which enabled them to develop empathy with each other and trust. A situation in which the game is played over and over is called an iterated prisoner's dilemma (IPD) and the outcome is very different from the basic one-off game. The payoffs are the same as in the one-off, but when neither side can deliver a fatal blow to the other and the opponents must meet each other repeatedly, a strategy is needed to maximize your longer-term payoff. But what should the strategy be? Could it lead to cooperation?

Political scientist Robert Axelrod asked this question in the late 1970s and invited scientists to submit solutions in the form of computer programs that would play different strategies against one another in a tournament that would decide the answer.[16] All kinds of complex strategies for winning the IPD were pitted against each other in the tournament, but consistently the best was the simplest of them all: Tit-for-Tat (TfT). TfT cooperates on the first move and thereafter copies the move of the opponent on the previous move. If the opponent attacked, TfT attacks; if they cooperated, TfT cooperates. The more complex strategies that came closest to matching the success of TfT did well because they cooperated with it, thus helping it. The message of the tournament was as clear and simple as the winning strategy itself: cooperation pays.

Conditions matching the TfT strategy can be seen in the exchanges between the two sides in the trench war. Soldiers often proposed cooperation to the other side; both sides cooperated when they could, but each retaliated if the opponent attacked. In the bloodiest conflict of modern times, the result was frequent cooperation across enemy lines. The reason that all this cooperation did not lead to peace is that it was continually destroyed by the opposing high commands who thought they could win and were careless at what human cost.

'Good-morning, good-morning!' the General said
When we met him last week on our way to the line.
Now the soldiers he smiled at are most of 'em dead
And we're cursing his staff for incompetent swine.
'He's a cheery old card,' grunted Harry to Jack
As they slogged up to Arras with rifle and pack.
But he did for them both by his plan of attack.

'The General' by Siegfried Sassoon

Situations such as the prisoner's dilemma crop up wherever two protagonists meet and they have the option whether or not to cooperate. The soldiers on the Western Front did not need to look up a textbook on game theory to work out what they should do to maximize their chance of getting out alive. Game theory hadn't been invented yet, for one thing. The point is that individuals following their own best interests can find themselves cooperating even with erstwhile adversaries. This is the first clue as to how selfishness, meaning simply doing what is best for oneself, can lead to cooperation.

A good deal of human cooperation arises through such enlightened self-interest, and yet it is difficult to see how it could explain the kind of selfless altruism shown by the *Topos* in the Mexico City earthquake. Could there be something deeper about the human propensity to cooperate? A biological impulse to cooperate? And if so, how could it have evolved? That is always the ultimate question we need to ask because evolution by natural selection got us here. How could natural selection, which favours any inherited trait that confers an advantage in survival and reproduction, cultivate an apparently selfless tendency to help others?

2

A river of glowing light

In St Petersburg, at four in the afternoon on 30 June 1876, a violinist stands at the open window of a small cottage, bow poised over the strings, waiting for a signal. When the nod arrives, it comes not with a conductor's baton but as an all-clear passed down a line of anarchist conspirators standing sentry along two miles of road stretching away into the distance. Each sentry is a lookout for anything that might obstruct the passage of a carriage passing at full tilt along the route. No lumbering peasant cart carrying a heavy load must impede the getaway. One sentry is set to walk up and down his beat carrying a handkerchief in his hand, another to sit on a stone by the wayside and to eat cherries. All are giving innocent-looking signals designed to avoid the attention of the Tsar's secret police.

At a little after 4 pm, the all-clear comes through and the violinist strikes up a wild mazurka. Across the road from the cottage is the yard of a prison hospital. On hearing the musical signal, a prisoner taking exercise in the yard sloughs off his heavy prison coat and starts to sprint for the open gate. He was informed of the escape plan only two hours earlier in a coded message passed to him concealed inside a pocket watch. A guard gives chase, hurling a bayonetted rifle at his head, but the prisoner, though enfeebled by two years of incarceration, stays just ahead and reaches the gate unscathed. He passes the soldier guarding it who happens to be turned away, deep in conversation with a friendly passer-by, another conspirator. It happens that they are talking about microscopy—the soldier used to work in the hospital laboratory. Out of the gate, the prisoner sees a waiting carriage, but he notices in alarm that the driver is wearing a military cap. Could this be a trap? The escaped prisoner claps his hands to attract attention and as the driver turns, the prisoner recognizes a friend. The driver recognizes that the prisoner is Prince Peter Kropotkin (Figure 3).

Figure 3 Peter Kropotkin

That evening, all of St Petersburg is in uproar. The Tsar himself commands that Kropotkin must be found immediately, but Kropotkin's friends have everything arranged. Shaved of his beard and dressed in top hat and tails, the prince hides in plain sight, dining with his friends at a fashionable restaurant that the secret police haven't the wit to search. With wanted posters everywhere and all his friends being interrogated by the secret police, Kropotkin cannot stay in Russia. Using a friend's passport and disguised in military uniform, he escapes through Finland to Sweden and from there via Norway to Britain, already the refuge of Karl Marx and other political exiles.

Prince Peter Kropotkin's escape from prison in Russia is as clear a demonstration as any of teamwork and altruism. Twenty people were involved in the plot to free him and all risked sharing his own fate of imprisonment, exile, or worse.[1] Dramatic though the story is, Kropotkin has an even better claim to a role in the history of cooperation through the book that he later wrote while in exile in Britain, *Mutual Aid*.

Prince Peter Kropotkin (1842–1921) was born into the highest echelons of the Russian aristocracy. His father was a general and an owner of more than 1,200 serfs, but Peter was raised in the compassionate care of serf servants whose gentle, open-heartedness left a lifelong impression upon his character.[2] At the age of 20, Kropotkin rejected military service at the Tsar's court and instead applied for a commission in a Cossack regiment in Siberia, thinking that in this wild region he could give free rein to his interest in natural history and geography. Primed by reading Darwin's recently published *Origin of Species*, he arrived in Siberia expecting to see competition everywhere among the wild animals he encountered. Instead, what struck him was how dependent animals were upon other members of their species, not only in mating but also in defence and feeding. The way in which ants and bees cooperate within a colony was well known, but there were other examples too. For example, normally solitary burying beetles cooperate with their mates and sometimes others to bury the corpses of small animals as food for their young.[3]

Kropotkin spent five years in Siberia, during which time he made and published important scientific observations, but he also encountered the stubborn opposition of the Russian state towards improving living conditions for people in the region. The experience contributed to convincing him that justice for the masses could only come by abolishing the state altogether and replacing it by spontaneously organized cooperation. Agitating for these anarchist ideas back in St Petersburg, he was arrested and imprisoned in the dread fortress of St Peter and St Paul. He made his dramatic escape two years later.

Russian evolutionists who, like Kropotkin, were familiar with the extreme rigours of the Russian climate and its destructive impact upon living things tended to interpret the Darwinian struggle for existence as an endurance test against physical forces.[4] However, in the more crowded land and less exacting climate of England, Kropotkin found that scientists such as Thomas Henry Huxley and others regarded the struggle for existence as a competition for resources among members of a species, rather than as a struggle against the physical environment. The English world view was heavily influenced by the Rev. Thomas Malthus's book *On the Principle of Population*, which argued that the human population must always increase to the limit of available food resources, whereupon nature will take its fatal course.[5] Malthus's book inspired Charles Darwin and Alfred Russel Wallace, both of whom appreciated that its principle should apply to all species in nature.

They independently realized that natural selection would favour the spread through a population of any inherited characteristic that gave its carrier an edge in the struggle for existence, and that this would lead to evolutionary change in species.

On reading an essay by Huxley on *The Struggle for Existence in Human Society*, Kropotkin, now established as a writer and journalist in England, recoiled from its Hobbesian brutality. Huxley wrote:

> [It] could be said of ancient man, in his 'savage' state: the weakest and the stupidest went to the wall, while the toughest and the shrewdest, those who were best fitted to cope with their circumstances, but not the best in any other way, survived. Life was a continuous free fight, and beyond the limited and temporary relations of the family, the Hobbesian war of each against all was the normal state of existence.[6]

Or in other words, society emerges in opposition to natural selection, not from it. Predisposed by his Russian education, by his first-hand observations in Siberia, and undoubtedly by his anarchist views, Kropotkin began to write a series of articles on cooperation and evolution in answer to Huxley. These first appeared in the influential periodical *The Nineteenth Century* and were then collected together in his book *Mutual Aid, a Factor in Evolution*, published in 1902.[7] The book is a classic text, full of examples of cooperation and social organization among animals and a long, historical account of mutual aid in human history. The marrying of animal and human examples within a single volume is strongly reminiscent of Darwin's later books that emphasize the continuity between the two.

Kropotkin spent 40 years in exile in western Europe, most of the time in England, but also undertaking several extensive speaking tours in the United States. He was an ailing old man by the time the hated Tsarist regime was overthrown in the February Revolution of 1917, but he could not stay away from what seemed like the realization of his lifelong dream of freedom for the Russian people. He left Britain under an assumed identity, but word got out and at each stage of his journey east the now famous writer and radical was welcomed by enthusiastic crowds. As his train pulled into Petrograd station at 2 am at the end of his homecoming journey, he was met by military bands playing the Marseillaise and 60,000 people waiting to welcome him.

The promise of the revolution was not to be fulfilled. In the power struggle that followed, the Bolsheviks led by Lenin achieved absolute power. Kropotkin wrote scathingly: 'Revolutionaries have had ideals. Lenin has none ... Things called good and things called evil are equally meaningless to him.' Anarchists, and other political opponents of the Bolsheviks, were assassinated, but Kropotkin was spared, probably because it was obvious he would not live long and his murder would cause an international outcry. He died near Moscow in February 1921. The English poet and anarchist Herbert Read later wrote of his death:

> A river of glowing light
> poured into the open grave
> all the light in the world
> sank with his coffin
> into the Russian earth.

From 'The Death of Kropotkin' by Herbert Read, 1950

Kropotkin idolized Darwin and believed that mutual aid among members of a species was fully consistent with the latter's theory. While that is true, Kropotkin had overlooked something important. Natural selection favours the individuals that are best adapted, which thereby leave more offspring. Kropotkin's argument that mutual aid would evolve out of the struggle for existence was based upon the advantage that mutual aid confers upon the group, not the individual. Kropotkin recognized the difference, but he did not see that he had undermined the explanatory power of natural selection by changing its basis from the individual to the group.[8] Many others have since fallen into the same trap, believing for example that natural selection operates for the good of the species. It doesn't and it's not hard to see why that is from a deeper examination of an example that appears in *Mutual Aid*.

Since Peter Kropotkin described beetles teaming up to bury corpses for their offspring to feed on, we have learned much about what happens among this society of gravediggers. This research clearly shows how actions that may appear to benefit the group only occur when there is benefit for the individual. As Kropotkin believed, beetles cooperate because a single beetle cannot easily bury a corpse by itself, and there is an additional imperative: it has to be done quickly. Carrion is a resource that is prized by crows, foxes, and other animals; flies are always ready to lay eggs and infest a corpse with maggots; a dead body is an all-you-can-eat buffet for bacteria and fungi. Defence of a valuable resource against so many competitors can be a powerful incentive for burying beetles to cooperate with each other.

As the beetles excavate the soil beneath the dead animal to bury and hide it, they also strip it of feathers and fur, fashioning the naked corpse into a ball and applying a coating. This coating is secreted from the beetle anus, inoculating it with yeast and bacteria from the beetle's own gut that inhibit the growth of bacteria that would quickly turn the corpse putrid.[9] The coating not only preserves the flesh of the corpse but also stops the smell of putrefaction that attracts flies. The beetles clean the corpse of any fly eggs that do appear. If a corpse is already too maggot-infested, burying beetles show less interest in trying to colonize it.[10]

The embalmed corpse becomes a subterranean, edible nest on which several families of beetle larvae may be collectively reared (Plate 1). The youngest larvae are fed pieces of carcass by adults of both sexes in the nest. This sounds like the very picture of beetle domestic bliss, albeit from the Addams Family, but cooperation among adults caring for the nursery inhabitants is not unconditional. Cooperation only occurs when sharing parental care is to the advantage of the individual family, or when beetles cannot tell which are their own larvae and which belong to another beetle family.

When a carcass is small enough to be buried by a single pair of burying beetles, it does not share the resource and will fight off other beetles that attempt to lay their eggs there. But there is another situation in which otherwise hostile burying beetles cooperate with each other: the presence of competition. Carcasses being consumed by fly maggots emit a characteristic foul smell due to a compound called dimethyl disulphide (DMDS). Exposure to DMDS alone is enough to induce burying beetles to cooperate with each other.[11] To a burying beetle, DMDS is a loud siren of alarm that warns of the need to cooperate in the face of a common enemy.

The burying beetles illustrate one of the simplest situations in which cooperation evolves under natural selection: direct benefit to the cooperators. Once we understand the natural history details, it is clear that beetles only cooperate when it is to their individual advantage to do so. Just as for soldiers in the trenches, cooperation among burying beetles is conditional. There is no sign of altruism here. Furthermore, a hypothetical gene that caused a burying beetle to help another mother without laying any eggs of her own in the nest would obviously not transmit that gene for altruism. This is the problem in explaining the evolution of altruism: it is a dead end.

Burying beetles are not *Topos*, but we know for example that altruism and cooperating without a direct benefit are real phenomena, not least in our own species. Among social bees and ants, there are sterile female workers that spend all their short lives gathering food for a queen who is the only individual to reproduce. Kropotkin was right that the human and animal worlds are full of examples of mutual aid, even if he was unable to square the most extreme examples with natural selection. Can we solve this problem another way?

3

From selfish genes to social beings

Mariners have to cooperate with one another. It's the only way to survive in a hazardous environment. This is especially true if you live the brutal life of a pirate. In the so-called Golden Age of piracy of the late seventeenth and early eighteenth century, the crews of pirate ships were as thick as thieves. Crews of over a hundred were not uncommon. Blackbeard's crew was 300 men.[1] How do you obtain cooperation and avoid conflict in such a large gang confined on a small wooden craft? The answer is interesting because it contradicts what one might expect. After all, who could be more selfish or barbarous than a pirate?

The crews of the merchant ships that were the prey of pirates were ruled by fear and corporal punishment. A merchant captain could do practically what he liked to control his crew and frequently did. But, by contrast, the violence of pirates was mainly directed outwards, rather than upon each other. Pirate ships were stolen vessels, owned by the whole crew. The pirate captain was a first among equals, with the same accommodation and fare as the men—female pirates existed but were very few. The captain was democratically elected by the crew who would not infrequently dismiss and even punish him for offences such as cowardice or lining his own pockets. To further control abuse there was a separation of powers on board pirate ships before ever the device was adopted by democratic governments. When not actually in battle, power over discipline was exercised not by the captain but by a quartermaster who was also elected. Conditions for the crew of a pirate ship were so much better than those experienced by the sailors on the merchant ships they raided that those ships were a source of new pirate volunteers as well as booty.

Pirate ships not only practised a separation of powers in their governance but also had a constitution or set of articles that guaranteed every pirate the right to vote, a fair share of food, drink, and plunder, and compensation in case of disability. There were also disciplinary rules while on board: no women, no gambling, no naked flames around the gunpowder store, no fighting, and lights out at 8 pm. Punishments for breaches of the rules could be severe but were not arbitrary, as they could be on merchant vessels. Pirate constitutions were deliberately framed to further cooperation and foster fairness among the crew, or as one court disapprovingly said, pirates were 'wickedly united, and articled together'.[2]

The anarchist Peter Kropotkin, had he known how pirates governed themselves, would undoubtedly have called it mutual aid. Merchant vessels and pirate ships exemplify the two alternative ways to obtain the cooperation necessary to operate successfully at sea: coercion or community of interest. What is perhaps surprising is that pirates were the ones motivated by community of interest.

Pirates were not the most dangerous cooperators on the high seas, even before bedtime at 8 pm. That prize goes to another predator that still operates as a wickedly united team: the Portuguese Man O'War (Plate 2). This creature resembles a jellyfish but belongs to a related group called the Siphonophora. Siphonophores are not so much individuals as colonies or floating villages of specialized individuals called zooids, all descendants of a single founding ancestor. The community of interest of the zooid team lies in their shared genes rather than the pirate code, but otherwise the teams are remarkably similar.

At the top of the colony is a zooid modified to form a gas-filled bladder that keeps the whole colony afloat.[3] The bladder bears a tall membranous crest fringed with a purple rim that acts like a sail, propelling the colony along in the wind. Strangely, Portuguese Man O'War display handedness—half have a left-handed sail and half a right-handed one, the difference being determined during early embryonic development. The wind propels left and right-handed sails in opposite directions. Below the bladder, budding into the surface of the sea, are attached other kinds of zooids with specialized functions.

Gastrozoids have a mouth and the ability to secrete digestive enzymes that break down food. Gastrozoids have no means of catching food for themselves but rely for this on another specialist zooid that has a tentacle but no mouth. Each tentacle, which can be as much as 30 metres long, is generously equipped with stinging cells that on contact with a fish or other soft prey inject it with a paralysing toxin. Once trapped, the tentacles contract like the coils of a spring, hauling the prey up to the surface and into the waiting mouths of the gastrozoids. The moment an item of prey has been captured, the gastrozoids begin writhing with mouths open like ravenous gorgons. Fifty have been observed latching on to a fish 10 centimetres long, completely engulfing it with their mouths. The enzymes released by the gastrozoids quickly turn the prey into a soup that is released into a communal cavity where all zooids can share the food.

Sex in the Portuguese Man O'War is also a delegated function, falling to the responsibility of another specialized zooid that produces sperm or eggs, depending on the sex of the colony. The reproductive zooids cluster inside a structure that, if this were a space mission, would certainly be called a landing module because it separates from the colony when mature. This module contains not only reproductive zooids but also small tentacle-bearing zooids, some of those mouthy gastrozoids, zooids modified for propulsion, and some jelly polyps. No one knows what the jelly polyps do, but then every big team has a role like that, doesn't it?

If we were to draw up a pirate-style list of the articles of association for a Portuguese Man O'War, it might look something like this:

1. We go where the wind blows—left or right, Captain Bladder decides.
2. All zooids are equal owners of the genome.
3. All prey is to be shared.
4. No sex on board.

Cooperation among pirates on a ship and among the zooids in a Portuguese Man O'War is founded upon exactly the same principle: a community of interest. The pirate crew were equal owners in their ship and its plunder. Zooids in a colony are all clones, sharing all their genes and thus having an equal interest in the success of the colony, whether they are specialized for buoyancy, fishing, eating, propulsion, sex, or whatever it is jelly polyps do. The big question is: how much cooperation in other organisms can be explained this way?

Among land animals, a colony structure resembling that found in the Portuguese Man O'War can be found in bees, wasps, and ants—the social Hymenoptera. In the European honeybee for example, the caste of sterile female worker bees is dedicated to feeding and caring for the colony, while egg production is the exclusive preserve of the queen. Males (drones) just mate. Ants have a similar social structure, with the addition in some species of a soldier caste that specializes in defence.

If drones are the emblem of male idleness, then worker bees are the Victorian embodiment of female virtue, labouring all day for the good of the hive and never indulging in sex. From a biological perspective, a worker bee seems to be the ultimate altruist, bearing only costs and no offspring. This kind of family structure, with a sterile caste that helps to raise or defend a relative's offspring, is called 'eusocial'. Eusocial organisms are teams that transmit their genes as a unit.

How can eusociality evolve? The difficulty is that natural selection favours genes for behaviours that increase the representation of those genes in future generations. But altruists do the reverse, helping others at a cost to themselves. Explaining self-sacrifice was, Charles Darwin wrote, 'by far the most serious special difficulty, which my theory has encountered'. How can beings that leave no offspring evolve in the first place? Darwin suggested an answer.

He observed that all members of a social insect colony belong to the same family so, in modern parlance, they share genes. The behaviour of worker bees is commanded by shared genes that are transmitted by another team member specialized for reproduction: the queen. It's easy to see how this works when, as in siphonophores, all members of the colony are genetically identical. Social Hymenoptera are similar to siphonophores in that they live in colonies comprising specialized individuals, most of which are sterile, but there is an important difference. While all zooids in a colony are genetically identical, insects in a nest are not.

Worker bees are sisters, sharing on average half their genes with each other and with the queen, their mother. This means that half a worker bee's efforts benefit genes that she does not carry herself. So, was Darwin wrong? The gist of the answer was first provided in a throwaway remark by the British biologist J. B. S. Haldane, who quipped that he would be prepared to lay down his life for two brothers or eight cousins. Brothers, or more generally siblings with the same two parents (full sibs), share half their genes with each other. So, if one brother sacrifices himself to save the lives of two others, in purely genetic terms it is as if he had saved himself. First cousins share one-eighth of their genes, hence Haldane's joke that he'd need to save eight of them to justify his own sacrifice. We'll see later that Haldane was an extreme risk-taker who actually endangered his life to help others without pausing to calculate the cost to himself, but for now let's stick with the genes.

It was a decade before Haldane's insightful joke, made to students in a pub, was translated into a testable theory of social evolution capable of explaining what goes on in animal societies, and then two scientists came along with it at once. In 1964, William D. Hamilton defined the concept of *inclusive fitness* for the collective benefits of social behaviour among relatives. The simple rule he came up with was that a gene for helping a relative will spread if the cost to the helper is less than the benefit to the recipient multiplied by the degree of relatedness.[4] So, for example, in Haldane's joke the cost to him is one life, the benefit to his brothers is two lives (one each) and they are related to him by 1/2. The two sides of the formula balance exactly: 1 × JBS Haldane = (2 brothers) × 1/2. It would have been an unusually bold student who replied to Haldane, 'Actually, Prof., that's not quite right', because for the brother-saving gene to spread, the cost needs to be *less than*, not equal to, the benefit.[5] This point is true of course, but it's the kind of wise-ass comment that gets you buying the next round of drinks in the pub.

Also in 1964, John Maynard Smith, who had studied genetics under Haldane and was the source of the pub anecdote, introduced the term *kin selection* for the kind of natural selection that Darwin had described operating on families. Maynard Smith showed that kin selection could explain the evolution of much social behaviour and that alternative explanations based on the 'good of the species' or the 'good of the group' could not.[6] We can easily see how kin selection works by applying Hamilton's rule to bees.

The genes that make worker bees sterile and self-sacrificing are transmitted by the queen of the hive, and by serving her, the workers are aiding the transmission of copies of their own genes through their mother. To see if this fits Hamilton's rule, we need to consider how the different castes in the beehive are related to each other. The queen mates with only one male so that workers, who are daughters of the queen, and all the larvae they care for are full sibs. Full sibs share half their genes with each other (that is, relatedness = 1/2). If workers were to mate and to have offspring themselves, these would also carry only half the mother worker's genes, the other half coming from the father. So, purely in terms of inclusive fitness, there is nothing to choose between a worker having her own offspring or instead helping to raise her sisters. But, if she can help more sisters than the number of offspring that she can raise herself, then childless helping is favoured by kin selection.

Note that Hamilton's rule is not just about relatedness within a social group but also about the relative costs and benefits of helping, measured in the number of offspring that result. This is where the benefits of a team come in. Recall that team membership can benefit individuals in two distinct ways: through force of numbers and through the division of labour. In social insects we see both at work, with huge numbers of workers serving a queen who is nothing more nor less than an egg factory. A queen honeybee can lay 200,000 eggs a year and some queen ants lay millions a month.[7]

The theory of kin selection predicts that the transition to eusociality seen in social Hymenoptera should only have taken place in groups where females are monogamous, so that all offspring in the family team are full sibs. Eusociality has evolved eight times independently in Hymenoptera and in all eight cases females were monogamous when eusociality arose, just as predicted.[8] There are also exceptions among eusocial Hymenoptera that prove this rule.

In some eusocial species, including common wasps (*Vespula* species) and honeybees (*Apis* species), originally monogamous queens later evolved a change to their mating habits and now mate with multiple males. However, in all these cases, workers have become genetically sterile, so they cannot switch to laying their own eggs even though this might theoretically be favoured by kin selection. Genetic sterility in workers permits queens to cheat on their helpers because workers cannot defect to raise their own offspring.

Beyond the social Hymenoptera, eusociality has evolved in a very miscellaneous roll call of animals consisting of some termites, gall aphids, a few thrips, a flatworm, an ambrosia beetle, snapping shrimp, and a handful of very weird mammals called mole rats. The naked mole rat and its other eusocial relatives belong to a group of rodents that is a sister to guinea pigs. Imagine a guinea pig shrunk to the size of a fat thumb, with no fur and pink wrinkled skin (Figure 4). Add very sharp incisors that protrude like buck teeth outside the lips, and you have the image of this odd animal. And its appearance is not the only exceptional thing about the naked mole rat. They live ten to twenty times longer than other rodents of similar size, reaching up to 30 years of age in captivity. Parents who give hamsters as pets to their children can expect to have to talk about death at an early date. Give a child a pet mole rat and you will need to discuss pensions.

Naked mole rats are native to east Africa, inhabiting semi-arid grasslands in Kenya, Ethiopia, and Somalia.[9] They live below ground in huge and

Figure 4 Naked mole rat

complex tunnel networks occupied by a colony of 75 or more closely
related animals of both sexes. Only one female in the colony produces pups,
usually mating with one male. The rest of the colony behave like workers,
cooperating closely with one another to dig the tunnels where they find
their food. A digging party is fronted by a lead animal using its incisors to
scythe away at the soil, which is passed back to a string of workers lined up
head-to-tail in the tunnel behind. These in turn sweep the spoil along to an
animal near the surface who kicks it over its head to where a molehill accu-
mulates. The purpose of all this subterranean activity, which can generate
kilometres of burrow, is to locate plant tubers, roots, and bulbs to supply
food for the whole colony. Their food is their source of water too, since
naked mole rats do not drink. (Don't let anyone tell you that this is why
they live so long.)

There is some debate among zoologists as to whether naked mole rats
and similar related species are genuinely eusocial, but the disagreements are
more semantic than biological, concerning exactly how 'eusocial' should be
defined.[10] So far as the animals themselves are concerned, the situation is
pretty clear. They live in big families consisting of a single breeding
pair, assisted by older offspring who are full sibs, and thus as closely related
to each other and to their parents as worker bees are to their queen

(relatedness = 1/2). Worker naked mole rats are physiologically capable of reproduction but don't reproduce within the colony. Monogamy combines with a peculiar set of circumstances in their natural history to favour family members who help raise siblings rather than breed themselves. Digging tunnels to find subterranean food is a cooperative enterprise that directly rewards staying in the family burrow. On the other hand, leaving home to find a mate is probably very risky, not least because naked mole rats are able to distinguish non-colony members by their call and will kill strangers.[11] Kin selection has done the maths and made naked mole rats eusocial.

While there are no eusocial birds, just under 10 per cent of species do breed cooperatively.[12] The main difference between eusociality and cooperative breeding is the degree of commitment by helpers to raising their siblings. In the former case, workers sacrifice any opportunity to breed themselves, while in the latter case family members spend some of their lives helping to raise their siblings before leaving to have families of their own. It has proved technically difficult to measure the costs and benefits of helping in cooperatively breeding birds, but the evidence collected so far does support Hamilton's rule. As expected, helping does increase the reproductive success of the breeders receiving assistance, and thereby provides some benefit to helpers through kin selection.[13] Those helpers also commit greater effort towards closer relatives.[14] But is the indirect benefit large enough to outweigh the costs of helping relatives?

This question has been answered directly for a population of long-tailed tits (Plate 3) studied in the city of Sheffield, England.[15] These small birds form monogamous pairs in spring and lay a clutch of about ten eggs. Nest predation in the study population was high, so that about 70 of the 100 nests in the area were destroyed each year. If this happened early in the breeding season the birds attempted to breed again, but if it occurred later, the males in particular started to feed young in another nest. Genetic testing showed that these helpers did not help at random but chose the nests of relatives for their visits with food. Broods that were close relatives of the helper received more help from them than broods that were more distantly related. The average relatedness between helpers and recipient broods was $r = 0.2$, which suggests a genetic relationship approximately like that between an uncle and a niece/nephew.[16] Relatedness is one of the three variables required to test Hamilton's rule. What of the other two variables: benefit and cost?

The benefit of helping was found to increase the breeding success of the recipient parent by about a third of an offspring. What can a long-tailed tit do with a third of a son or daughter, you may wonder? But evolution has done a lot with much less. Tiny improvements add up to large effects when multiplied over time. Looking at costs to the helper, he had already lost the opportunity to breed when he started helping, so there was nothing additional lost there and hence no immediate cost. However, the effort of helping did significantly reduce the survival of helpers into the following year when they would have a new opportunity to breed. This opportunity cost amounted to only 0.03 of an offspring, or about 10 per cent of the benefit. To complete the calculation of Hamilton's rule, we multiply the benefit of 0.3 by the relatedness 0.2 and get 0.06, which is twice the cost of 0.03. Hamilton rules!

Long before biologists took a professional interest in the social organization of animals, there were folk names for the groups they formed. A pride of lions, a murmuration of starlings, and a parliament of owls are collective nouns that express the perceived nobility of big cats, the awe-inspiring synchronized aerobatics of huge flocks of starlings, and the answering hoots of owls. Likewise, the collective noun for crows cast them with a reputation, but it was not a favourable one: a 'murder' of crows. In their black plumage with a bluish sheen, they even dress the ghoulish part. Crows (*Corvus corone*) are omnivores and opportunists that will very occasionally attack sheep during lambing, for which rare crime they were themselves murdered wholesale on upland farms in Britain.[17] We now know that these birds are intelligent and are sometimes cooperative breeders. Why they only sometimes cooperate is intriguing.

In most European populations, crows breed as a single pair and raise between one and five nestlings (Plate 4), feeding within a territory that they defend just during the breeding season. Four or five weeks after fledging, young birds leave their natal territory to forage more widely for food and it may be several years before they acquire a breeding territory of their own. In northern Spain, crows behave differently. Here, a breeding pair holds a territory all year round and is helped to rear its offspring by non-breeding young from previous broods and by immigrant males. In all, there may be as many as nine adults in the group, cooperating in raising the young of the breeding pair.[18] Bigger groups fledge more of the pair's offspring than they can on their own, so the advantage of cooperative breeding to parents is clear. But why do adult crows help raise another bird's offspring, delaying their own reproduction for up to four years, when they could be breeding themselves, and why does this happen only in northern Spain?

To the first question, there are two possible answers: there may be direct advantages to belonging to the group, such as access to a territory well supplied with food, and indirect benefits through inclusive fitness. Birds raised in richer territories were more likely to stay than birds raised in poorer ones. This effect of food supply was confirmed by the results of an experiment. Artificially increasing the food available in a territory led to young birds delaying dispersal longer than usual. There was also evidence that young birds that remained in their parents' territory learned from them where to find novel kinds of food. What of indirect benefits?

In some birds and mammals, older offspring assist with the rearing of their younger siblings. This is a behaviour that would clearly be favoured by kin selection in crows, as in other species. A comparative study of 18 cooperatively breeding birds and mammals found that helpers consistently discriminate in favour of kin and that this effect is strongest in cases where the benefits of helping are greatest.[19] In the northern Spanish crows, family members were not the only helpers: there were immigrants as well. Most of these birds were male and genetic analysis revealed that they were related to the male of the dominant pair. Thus, immigrant helpers were Dad's returning family and probably acquired indirect as well as direct benefit from helping him.

So, what appears to be altruistic helping among crows in northern Spain has both direct and indirect advantages for the helper. Rather than dispersing, the older offspring of a breeding pair stay for the home comforts and the advantages of group living and then join in with parenting for the indirect rewards of inclusive fitness. These direct and indirect rewards reinforce each other, accelerating the evolution of cooperative breeding compared to what would happen if only direct benefits or only kin selection were involved.[20]

The remaining question is why cooperative breeding in crows happens only in northern Spain. To check whether helping is determined by a genetic difference between crows in northern Spain and elsewhere in Europe, crow nestlings were experimentally swapped between nests in populations in northern Spain and Italy. If helping was an inherited trait, the transferred crows should behave like birds in their birth population, not their adoptive one. In fact, when fledged, crows showed the behaviour of their adopted population, not the one in which they were born. This suggests that crows respond flexibly to their social environment, with the likely deciding factor being whether the territory in which they are raised is defended year round or not. Where a territory is permanent, the advantage of staying for a young bird is obvious: a protected food supply in familiar surroundings. Frustratingly, we do not know why territories are held outside the breeding season in northern Spain but not elsewhere.

One might expect that permanent territory-holding would be favoured in harsh environments, which might be the explanation. Finding an answer to this question in crows will be difficult while families in northern Spain remain the solitary exception. However, there is a more general pattern seen across birds and mammals: in both groups cooperative breeding is more common in harsh environments such as deserts. The correlation of cooperation with harsh environments could be explained in two different ways: either animals evolve cooperative breeding when living in harsh environments, or animals that happen to be cooperative breeders colonize harsh environments because they do better there than other species that are not cooperative. Across groups as large as birds and mammals, we might see different explanations apply in different species, but in birds at least we have a clear answer.

A study examining the evolutionary ancestry of cooperatively breeding birds living in harsh environments found that as a rule they bred in this manner before they colonized from more benign habitats.[21] The same analysis showed that the colonization of harsh environments by cooperative breeders occurred at twice the rate of colonization by non-cooperators. A result that would make Kropotkin throw his fur hat in the air. Nice discovery though this is, it only tells us that cooperative breeding is an advantage if you already have the habit, not why cooperative breeding evolves in the first place. That is a different story.

We already know that successful cooperation requires an alignment of interest between members of the team. When the team is a family, there is scope for kin selection to operate in favour of cooperation, over and above any direct benefits. If this is how cooperative breeding evolves, then we'd expect it to be common in families where genetic relatedness is high and rare in families where relatedness is low. The range of relatedness within families found in nature is wide. High relatedness occurs where offspring are full sibs ($r = 0.5$), but it is lower when offspring share only one parent and have different fathers (half sibs, $r = 0.25$).

Where it has appeared, cooperation has evolved more often from ancestors that were monogamous than from those that were not, supporting the theory of kin selection for helping because the offspring of monogamous females are full sibs.[22] Cooperative breeding is also found among mammals, where it follows the same pattern found in birds. It has most often evolved in close family groups where the helpers are related to the breeding female and she is monogamous.[23]

Relatedness is not the only factor influencing the evolution of cooperative breeding because this also depends on the costs and benefits of helping, which vary with environmental factors. No cooperative breeders are known among seabirds, however monogamous they may be. Seabirds don't hold feeding territories because their food (fish) is abundant and mobile. In the absence of a territory supplying food there is no advantage for young seabirds in staying home after fledging.

The puffin nests in a burrow on sea cliffs and is entirely monogamous, which would make it a theoretical candidate for the evolution of cooperative breeding, but the breeding pair receive no help raising their solitary chick. A helper would have to be a bird raised in a previous season, and it's hard to see how such a helper could increase its inclusive fitness by helping a younger sibling that is already receiving the full attention of both parents. The helper would also pay a severe direct cost by losing a breeding opportunity for itself. In general, kin selection goes a long way towards explaining why animals cooperate and when they do not.[24] But cooperation is not confined to family groups. How can it pay to cooperate with non-relatives?

4

Big steak or big mistake?

The family motto was a single word: 'Suffer.'[1] It was displayed on the family coat of arms and engraved on the silverware used at the dining table. A dinner guest observed that the engraving had worn down with years of use and appeared to just read 'Supper'. The family with the worn silverware and the bleak maxim were the Haldanes. It was J. B. S. Haldane (Figure 5) who prefigured kin selection when he said that he would lay down his life for his relatives, provided enough of them were saved. Ironically, Haldane himself behaved altruistically towards others without calculating the risk to himself.

The Haldanes were landed gentry with a seat near Perth in Scotland. Since the family was moneyed, suffering was not forced on them by circumstance but chosen through a sense of duty and, in the case of J. B. S., an addiction to danger. His father, John Scott Haldane, was an Edinburgh-trained doctor who became a professor of physiology at Oxford, where J. B. S. was born. At the age of two, he was crawling around the floor of his father's home laboratory and at three he is said to have asked about the blood coming from a cut on his head, 'is it oxyhaemoglobin or carboxy-haemoglobin?' Apocryphal or not, the anecdote perfectly captures the scientific detachment that both father and son had towards pain. J. B. S. recollected later that his father disliked experimenting on animals and had trained himself to be indifferent to pain in order to use himself as an experimental subject. The object was 'to achieve knowledge that would save other men's lives'.

John Scott's interest in saving lives was not merely academic. He crawled into coal mines after disasters to investigate the explosive methane that accumulated, and he consulted for the Royal Navy on how to bring deep-sea divers safely back to the surface. His son would often accompany him on these trips. At age 15, J. B. S. spent time with his father off the coast of

Figure 5 J. B. S. Haldane

Scotland, experimenting with diving apparatus. Eager to participate in the experiments himself, Jack donned a diving suit and was lowered into the water. He had not been submerged for long before his suit began to fill with water through the ill-fitting wrist and leg bands that were designed for a full-grown man. J. B. S. stayed submerged, manipulating the gas valves in the suit to stop the water from rising above his chest until he was hauled from the sea, cold and soaking wet, at the end of the experiment.

At the start of the First World War in 1914, J. B. S. volunteered for the Black Watch, a Scottish regiment headquartered near the family seat. At the age of 22, J. B. S. found himself in the trenches on the Flanders battlefront. 'I was well aware that I might die in these flat featureless fields, and that a huge waste of human values was going on there. Nevertheless, I found the experience enjoyable, which most of my companions did not.' Haldane was made trench Mortar Officer and with a small band of men under his command he roved the trenches firing off primitive stove-pipe bombs at the enemy lines. The bombers were not popular with the other men because their activities invariably drew retaliatory artillery fire from the German lines, but by then the bombers had moved on. J. B. S. was in his element, making dangerous nightly forays across no man's land to listen in on what the German's were saying—he had learned German at the age of five from his nanny.

While in the trenches, J. B. S. somehow found the time and mental composure to complete a scientific paper on some genetic experiments in mice, following the death of a co-author in action.[2] Death and destruction were the constant companions of youth at that time. J. B. S. had several close calls and saved at least one comrade's life through extraordinary bravery, though it was another act of heedless self-sacrifice that saved thousands.

J. B. S. was briefly recalled behind the lines to assist his father who had arrived in France on an urgent mission. The Germans had started to use mustard gas and the Allied soldiers were being issued with a type of gas mask that John Scott Haldane had already warned the authorities was useless. Experiments with better respirators were needed to discover if they would work and allow soldiers to fight while wearing them. The experiments involved father and son breathing the deadly gas, which left volunteers bedridden for days and short of breath for months. The research led to better respirators and the blunting of mustard gas as a weapon. Characteristically, J. B. S. made light of the self-sacrifice involved in his contribution to this.

Thirty-five years later, the jocular reference that J. B. S. made to laying down his life for his kin came from a man who knew exactly what self-sacrifice meant. Haldane's joke is funny because of the incongruity between the very precise, algebraic conditions he gives for self-sacrifice and the fact that in such circumstances people never make a calculation, even when they know the odds. The very idea of a calculating hero is absurd.

Humans do favour their kin, but self-sacrificial actions such as those of J. B. S. and his father and the Mexican *Topos* are but extreme examples of normal human behaviour towards non-relatives. In the jargon of the behavioural sciences, we are 'prosocial'. Kin selection cannot explain why we are prosocial since kinship is not required. What other explanations can there be? The simplest hypothesis is that it is all a big mistake.

The Big Mistake Hypothesis (BMH) proposes that humans evolved in small family bands where kin selection favoured helping everyone around you.[3] With time, the size of human groups grew to include non-kin, but we still treated everyone as kin because this was baked into the human psyche. The BMH says that prosociality is not adaptive, it's an evolutionary hangover. Evolutionary biologists don't like hangovers. We dislike explanations like this because they explain away rather than attempting to explain. Nonetheless, the BMH needs to be taken seriously because anachronism is not that unusual in evolution.[4]

The human species in its anatomically modern form is only 200,000–300,000 years old and in all but the last 10,000 years of that history we lived in nomadic bands that hunted and gathered our food. Hence the argument often made that humans are ill-adapted for a modern lifestyle or diet and that we should behave or eat the way it is imagined hunter-gatherers did. This can lead to suspect claims based on what Marlene Zuk has called 'Paleofantasy'.[5] In the 1960s, Robert Ardrey popularized the idea that humans are innately aggressive and territorial because this was baked into our psyche during our evolutionary history.[6] Curiously, the BMH is the reverse of this argument, coming to the opposite conclusion that evolutionary history made us inherently prosocial. You wonder whether we could even be talking about the same species.

If there is evidence that supports the BMH, we should find it in our ancestry. We cannot directly tell how prosocial our ancestors were, but we can look at other primates with whom we share ancestors and work backwards. If we share a pattern that appears across all primates, then maybe it really is baked in. There are about 250 primate species and the scientific literature on primate behaviour is vast and inconsistent. However, one exceptional study managed to test 15 different species for prosociality with a standardized methodology.[7] Lemurs, chimps, spider monkeys, gibbons, macaques, human children, and a menagerie of other primates that would make Dr Doolittle drool were all tested in their normal social groups to see if they would spontaneously provide food to others when they themselves received no share. The result was startling.

As you might expect, not all primates were equally prosocial, but you might think that our closest primate relatives, the chimpanzees, would be most like us. In fact, on a scale of 0–10 for prosociality, humans were a 10 and chimps a 1. Why the big difference? Across the 15 different species, prosociality correlated closely with the extent of cooperative breeding. Species in which offspring received most care from others than their parents were the ones that most readily shared food without getting anything in return. 'It takes a village to raise a child' is not merely a proverb, as anyone who has raised children can testify. We humans are cooperative breeders, but chimps are not. We are prosocial, chimps are not.

This comparative study of primates suggests that prosociality does indeed originate in groups where kin selection operated, as testified by cooperative breeding. To this extent the study supports an assumption of the BMH. However, the BMH makes another prediction that needs to be tested, namely that prosociality in modern humans is not favoured by natural selection. Could that be true?

It has been argued that humans are too intelligent to blindly follow an injurious instinct against their better judgement. That is a matter of opinion. It would be better to test whether or not prosociality is of contemporary advantage. We'll return to humans anon, but first it is worth looking at the simpler case of cooperation among non-kin in other animals that lack the complicated baggage of a big brain.

Cooperation among non-kin is widespread among animals; it is even quite common in the social Hymenoptera.[8] Eusocial colonies evolve through kin selection, but in some species unrelated colonies help each other out. This seems to be most common in harsh environments in the early stages of colony formation when there are few workers to assist queens. Harvester ants live in the deserts of the American south-west where they can be found transporting seeds across the ground to their nests. Young queen ants have to do this work by themselves, which limits the rate at which they can produce workers to help them. A new nest might contain just ten workers, compared to the millions that inhabit some ant nests. In harvester ants, two unrelated queens may set up nests together, where workers and queens care for the brood of both. Experiments have shown that nests with two queens grow at a faster rate than nests with just one and that these colonies also survive better.[9]

Communal breeding as seen in harvester ants is distinct from cooperative breeding because the latter involves kin helping kin, while the former does not. Burying beetles, African lions, and acorn woodpeckers are among the many species of communal breeders. One of the chief benefits is better defence against predation, but as ever with cooperation it comes with an inherent threat from cheats—like false cooperators who dump their eggs and run.

A communally breeding bird called the greater ani has evolved an extraordinary system for identifying and ejecting the eggs of cheaters.[10] A nest is shared by two or three unrelated pairs of birds who chorus together in a complex and ritualized song that synchronizes egg laying and strengthens social bonds in the group. Some nifty egg shifting takes place to remove eggs laid too early and then communal brooding begins. Egg dumping by non-group females is common. Parents cannot recognize these foreign eggs as such, but they can tell whether an egg has been laid out of synchrony with the group from the colour of the shell (Plate 5). Greater ani eggs are covered in a dusting of a white mineral that wears off with time, revealing a blue shell beneath. Eggs that have just been dumped are therefore different in colour from those belonging to the community clutch and are ejected. The result is that cheating has low rewards and cooperative behaviour is protected.

Unrelated animals gain a variety of direct benefits by joining social groups. Aside from communal breeding, the most important is defence from predators. There is safety in numbers for an individual in a herd because vigilance is shared and the per capita risk of being attacked is lower than for a solitary animal. Two antelope are grazing peacefully on the plains of the Serengeti when they see that they are being stalked by a cheetah. 'Run!' says one antelope to the other. 'What's the point?', asks the other. 'No one can run faster than a cheetah!' 'I don't have to run faster than the cheetah', comes the reply. 'I only have to run faster than you.' In other words, there can be purely selfish reasons for being in a group.

The situation just imagined is a primitive game, in the formal sense, because the payoff (survival) for one antelope depends on the behaviour of the other. In this case the best strategy is to join a group and always run. What of more complicated situations such as the prisoner's dilemma? For example, animals feeding in a group split their time between watching for predators and eating. These are mutually exclusive activities, so time spent on guard duty is a cost with the benefits enjoyed by all. The temptation for any animal must be to keep its head down and just eat, leaving the guarding to other more socially minded suckers. If animals play tit-for-tat, meaning something like 'I'll take a turn at guarding if you do, but not if you don't', then cooperation occurs and can be sustained. This is the result of the IPD. But is this how animals actually cooperate? Do they reciprocate?

Decades of research have gone into answering this question and a fair summary of the result would be: rarely.[11] We now understand much better how cooperation works in nature, including in our own species. The first lesson is that the dilemma as originally posed is a real one: when cooperation bears a cost, cheating can and does happen and there has to be a way to control this or cooperators will all defect. The IPD makes some assumptions about how the game is played, such as: it is divided into rounds, there are just two players, and there is no communication about intentions. When these are satisfied, some animals do play tit-for-tat as expected.

For example, lab experiments with Norway rats were run to test whether they would behave reciprocally in helping each other to reach food inside their cage—they did. In another experiment designed to reward or punish behaviours according to the payoffs expected in the IPD, rats cooperated in a tit-for-tat manner as predicted.[12] The problem with generalizing from these experiments is that they were run in very artificial conditions. When rats in this kind of experiment are hungry, they make ultrasonic begging noises, and another rat hearing this provides help in proportion to the intensity of the calls.[13] Hungrier rats get more help. This is not in the tit-for-tat strategy and shows that rats have a higher propensity to help each other than you would expect if they simply reciprocated. In lab experiments of similar design, humans also quickly cotton on to the rules of the IPD and play the tit-for-tat strategy. However, here too adding a little realism upsets the result. If participants in a test are allowed some time to talk to each other before the experiment starts, they become able to predict whether the other player will cheat, and they play accordingly.[14] A fundamental flaw in using the IPD to model cooperative behaviour is that social animals communicate with one another and the IPD ignores this.[15]

Intelligent animals do reciprocate, but cooperation among non-relatives is not solely dependent on reciprocation as the IPD would predict. Prosocial behaviour in humans goes further because it seems to be the default response in many circumstances; it is not dependent on reciprocation, nor is it confined to people who know each other. Furthermore, the benefits of helping are often unknowable and so cannot enter into a calculation as to whether or not it is worth the cost. We are daily considerate towards people whom we have never seen before and may never see again. And if you are thinking at this point of an annoying instance when someone was inconsiderate, ask yourself why you are so annoyed about it. Isn't it annoying because a norm has been broken? Anti-social behaviour provides the exception that proves the existence of a prosocial rule. Prosociality is a strategy for cooperating in a large group.

As far as humans are concerned, we are back to the question of why we help each other as much as we do. And what stops anti-social behaviour disrupting prosociality, turning us all into disenchanted versions of Gordon Gekko, who believed that 'greed is good'? You will recognize that these are essentially the same two questions that we have returned to repeatedly.

How does the BMH measure up as an answer so far? Our brief tour of cooperation between non-kin among animals suggests that there are many kinds of advantage to prosociality that could also occur in humans. One of the most important is seen in how we raise children.[16]

Sarah Blaffer Hrdy says in her book *Mothers and Others* that:

> It was the end of the twentieth century before evolutionary anthropologists like myself began to consider just how hard it would have been for foragers to rear surviving children, and then piece together disparate strands of evidence indicating that the help of group members in addition to the genetic parents was absolutely essential for the survival of infants (birth to weaning) and children (weaning to nutritional independence) in the Pleistocene.[17]

The evidence she assembled in *Mothers and Others* is persuasive. To begin with, human mothers are unusual in their readiness to let others hold their babies. Chimp and monkey mothers never allow others near their newborn infants and hold them close at all times. In the Hadza tribe of northern Tanzania, one of the last remaining hunter-gatherer communities still leading a traditional way of life and the go-to population for evolutionary anthropologists, newborns are held by others 85 per cent of the time. The same pattern is seen among nomadic Aka and Efe communities in Central Africa where 25–30 mothers will share their babies with each other, caring for and nursing them. Efe babies average 14 different caregivers in the first days of life, including fathers, brothers, and older female relatives. Shared suckling is not observed among apes but occurs in the large majority of human foraging societies studied by anthropologists throughout the world. The food sharing that begins at birth continues throughout childhood.

Though relatives are involved in the care of infants and children, non-relatives are important too. Bands of hunter-gatherers tend to be fluid in composition and relatedness among their members is typically low,[18] reinforcing the point that kin selection alone does not account for who helps whom and why. Cooperative breeding almost certainly provided the seed-bed for prosociality in humans, but this evolutionary development probably occurred in the ancestors of *Homo sapiens*, which would mean we have always been prosocial. One reason for pushing back the origin of prosociality to a period well before the origin of our own species is that recent research suggests that a second assumption of the BMH is a big mistake in itself. Not only is relatedness in bands of hunter-gathers low when the BMH assumes it is high, but there is archaeological evidence that in pre-modern times hunter-gathers cooperated in large numbers.[19] Over the hundred years

Figure 6 Bison

or so that anthropologists have been interested in studying hunter-gatherers, they have lived in dwindling, marginalized communities who could not muster large numbers. But the structures that hunter-gathers from previous times left behind them tell us that this was not always the case.

The Great Plains of North America stretching from Texas in the south to the province of Alberta in Canada were once roamed by huge herds of bison (Figure 6) that grazed in scattered groups across square miles of grass-land in a perpetual peregrination. Tens of millions of these huge beasts, weighing up to one-and-a-quarter tonnes, inhabited the Great Plains before the arrival of Europeans and the near extinction of the bison which fol-lowed. The wanton slaughter of bison (often called buffalo) was at least in part a military policy by the United States directed at starving the Native Americans, in particular the implacable Sioux, into submission. 'Kill every buffalo you can. Every buffalo dead is an Indian gone', was the view of one Colonel Dodge.[20]

A later Sioux elder named John Fire Lame Deer (1903–1976) told his biographer:

> The buffalo gave us everything we needed. Without it we were nothing. Our tipis were made of his skin. His hide was our bed, our blanket, our winter coat. It was our drum, throbbing through the night, alive, holy. Out of his skin we

made our water bags. His flesh strengthened us, became flesh of our flesh. Not the smallest part of it was wasted. His stomach, a red-hot stone dropped into it, became our soup kettle. His horns were our spoons, the bones our knives, our women's awls and needles. Out of his sinews we made our bowstrings and thread. His ribs were fashioned into sleds for our children, his hoofs became rattles. His mighty skull, with the pipe leaning against it, was our sacred altar. The name of the greatest of all Sioux was Tatanka Iyotake—Sitting Bull. When you killed off the buffalo you also killed the Indian—the real, natural, 'wild' Indian.[21]

When Europeans reached the Great Plains, the plains tribes were mounted on horses and were formidable warriors, but horses were relatively new to them. Although the horse had evolved in North America, it went extinct there along with other megafauna about 10,000 years ago. The horses that Native Americans were riding when Europeans encountered them were the descendants of escaped Spanish horses brought to North America when the conquistadores arrived in Mexico.[22] European accounts of native bison hunting describe it taking place on horseback, but the traditional method was on foot and required large numbers of people to cooperate. Archaeological evidence of this has been uncovered in hundreds of places, but most spectacularly at a site called Head-Smashed-In in southern Alberta.

Head-Smashed-In is a buffalo jump where herds of one or two hundred animals at a time were driven pell-mell over a sandstone bluff to die on the rocks beneath. Debris beneath the cliff has been excavated to a depth of ten metres, revealing that the site was in use for 6,000 years until the last occasion, in the mid-1800s. One hundred thousand animals may have perished at Head-Smashed-In, each big animal yielding 400 kg of meat, plus hides, bones, and other useful parts. The site must have been carefully chosen by the Native Americans who first used it because the landscape behind the jump has natural and artificial features perfectly adapted for the purpose of driving herds towards their doom.[23]

Four kilometres behind the bluff is a wide, 10 km long natural depression or gathering basin into which bison were carefully ushered by tribespeople sheltering behind long drive lines of stones piled into cairns. One such drive line made by the Athabasca tribe is 50 km in length.[24] Once gathered in the basin, the herd was lured and chased in the direction of the cliff by young men disguised with buffalo or wolf skins. Further drive lines sheltering more helpers funnelled animals from the gathering basin towards the killing ground. Care and coordination would have been needed by the scarers to stay out of sight until the herd was correctly positioned, or the bison would move in the wrong direction. If cornered, the huge animals might turn in a moment upon their tormentors and attack with horns tossing. Even with a big steak in prospect, no one wants to risk being kebabbed before dinner.

The nomadic Native American tribes travelled in groups of about 70 people, but when a buffalo drive was planned, they would gather in the hundreds or even thousands to participate. Large numbers of people of all ages were needed, not only to build and maintain the drive lines and create hides of brushwood around the cairns but also to shoo the animals along. Once over the cliff, there would have been mayhem with animals thrashing in their death throes and blood everywhere. Five thousand arrow heads have been recovered from the site, indicating that bows and arrows would have been used to finish the animals off from a safe distance. Then stone tools would have been used to skin and dismember the carcasses, strip the meat, sever sinews, and gather the harvest as quickly as possible before it began to decay. All the while, wolves would have been howling nearby and eagles wheeling in circles overhead, eager to seize a share.

The huge communal harvesting of meat practised at Head-Smashed-In appears typical of how nomadic hunter-gatherers cooperated in large numbers on every inhabited continent.[25] It is quite different to any scenario envisaged in the BMH, and suggestive of another form of cooperation entirely: interdependence. The Interdependence hypothesis proposes that the fitness of each member of a group is dependent to a significant degree upon the assistance of others.[26] Cooperation is required to hunt large animals, but the result is meat and material resources in quantities beyond the wildest dreams of a lone hunter. The cooperative rewards of a big steak give the lie to the idea that prosociality might be a big mistake. When cooperation yields a superabundance of perishable food, sharing the spoils with everyone including strangers is a costless means of building relationships on which you can depend. Surely this is why we are so willing to share food with strangers when no other animal so readily will.

If interdependence based on cooperative child rearing and hunting is the evolutionary basis of prosociality, it is older than *Homo sapiens*. We can be reasonably sure of this because *Homo erectus* got there first. *Homo erectus*, who appeared 1.8 million years ago, was born in Africa like our species, and also like us they migrated out of the home continent and colonized as far east as China. Wherever they are found, the fossil camps of this ancestor of ours are littered with elephant bones, and whenever *Homo erectus* showed up the local elephants disappeared soon afterwards.[27] *Homo erectus* liked nothing more than an elephant steak. The only way great apes could kill a herd animal the size of an elephant is through cooperative hunting. And as we have seen, this would have produced the original all-you-can-eat buffet.

The abundance of food produced by cooperative hunting is an example of what economists call a 'public good'. Public goods are resources produced by the collective efforts of individuals in the community and from which all may benefit, whether they contributed to those efforts or not. As ever, we need to ask what stops free riders taking advantage of everyone else's hard work when a public good is available to all? What's to stop a free rider just hunkering down in safety behind a cairn on a soft bed of hay to take a nap and then rocking up to the feast after all the dangerous hullabaloo is over? It's clear what won't stop this: reciprocity. If a chief is calling all the tribes together for a massive buffalo feast, he can't say, 'Either everyone participates, or the whole thing is off!' That way, you are giving free riders a veto and no cooperative venture will ever get off the ground. To maintain cooperation in the production of public goods, something more effective than crude reciprocity is needed.

An alternative strategy that supports cooperation in public goods situations is for cooperators to selectively team up with each other, leaving non-cooperators out in the cold. When Hadza people decide who they want to camp with they perform this kind of selection. As a result, camps differentiate into some that are more cooperative and others that are less so.[28] Online social experiments designed to test how people share public goods within a social network show the same phenomenon. Cooperators find and stick with each other, but non-cooperators are unfriended, and they experience this as punishment.[29] Adults, children, and some animals will go out of their way to punish non-cooperators, thereby defending and sustaining cooperation.[30]

Having a good reputation is important to attracting cooperators and acquiring the benefits, which is why we are so intensely interested in what others think of us. Charles Darwin suggested this in *The Descent of Man*,[31] but it was an idea already long in common currency. No one can express it better than in the speech that Shakespeare puts into the mouth of the duplicitous Iago in the play *Othello*:

> Good name in man and woman, dear my lord,
> Is the immediate jewel of their souls.

Who steals my purse steals trash, 'tis something, nothing:
'Twas mine, 'tis his, and has been slave to thousands:
But he that filches from me my good name
Robs me of that which not enriches him
And makes me poor indeed.[32]

Reputation, fairness, and trust all matter and are all a means of securing the rewards of cooperation and interdependence. This interdependence produces group-level behaviour, but it does so through benefits to the individual members of the group. It is difficult for cooperation to evolve through benefit to the group as a whole as envisaged by Kropotkin in *Mutual Aid* because, if there are no individual benefits to be had, liars and cheats prosper, fatally undermining the cooperation of the group.

Cooperation over sharing public goods may be favoured by interdependence, but rivalry is always there, just beneath the surface, and there are rival theories about how it can be kept in check. The first is not so much a theory of cooperation as a diagnosis of its failure made by an ecologist named Garrett Hardin. In 1968, Hardin published a brief but hugely influential paper entitled 'Tragedy of the Commons', in which he rehearsed the theory of Thomas Malthus that human population increase must inevitably outstrip the supply of resources. Hardin's diagnosis was that the Malthusian problem arises because, when resources are shared in a common pool, every individual is motivated to take as much as they can. This is a form of prisoner's dilemma, in which cooperation between rivals is rendered impossible by the fear of losing out.

For example, it might make sense to manage a resource such as a deep-sea fishery sustainably, but this requires self-restraint on the part of rival fishers, which is just not a winning strategy in a free-for-all. As a result, virtually all ocean fisheries are currently overfished and headed for commercial extinction. Hardin argued that there is no solution to such tragedies of the commons without external intervention. His preferred remedy was to privatize common resources on the grounds that it is in an owner's self-interest to conserve resources rather than to overexploit them. An alternative might be to impose top-down rules on the commons, which requires an institution such as a Hobbesian strong state.

Hardin's 'Tragedy of the Commons' framed the environmental debate that developed in the 1970s and to this day lies behind strategies for environmental protection such as debt-for-nature swaps that aim to privatize natural areas such as rainforest in the Global South.[33] Hardin's argument mesmerized the environmental movement with the seemingly unanswer-able logic of Malthus. However, a logical argument is only as good as the premises on which it is based and Hardin's premise, like Malthus' before him, was flawed. The challenge to the argument in the tragedy of the com-mons came from economist Elinor Ostrom (Figure 7) who scrutinized the way that commoners who share resources such as grazing land and irriga-tion water actually behave.[34] What she discovered was that commoners negotiate shares among themselves and create rules and institutions to pre-serve their individual and collective interests. Ostrom was awarded the Nobel Prize in Economics for her work, the first female recipient of the prize.

Ostrom found that cooperation emerges among people sharing a com-mon pool resource when they are able to communicate and work with their neighbours over time, building up reputation and trust. In the Swiss alps, farmers have for centuries grazed their cattle in the glorious alpine meadows during summer. Overgrazing is prevented and the grazing is

Figure 7 Elinor Ostrom

shared according to ancient rights that forbid anyone from pasturing more animals than they can feed over winter. Peter Kropotkin described cooperation among Swiss mountain villagers in his book *Mutual Aid*. Also, in Japan, common lands in the mountains have been regulated for hundreds of years by local village committees that set rules to ensure each family does a fair share of the work needed to manage the land and receives a commensurate share of the resources in return. Each village has a written code specifying the penalties for breaking the rules and most villages employ a detective who patrols on horseback. Fines for breaking the rules can range from payments made in saké to financial donations to the village school.

A characteristic of all cases of successful, long-term sharing of common pool resources is that they are regulated by institutions that are democratic and local. Rules are set by the community itself and enforced by graduated sanctions on those who break them. People in each community have a history of working together stretching back generations and expect this to extend into the future. Shared festivities and culture in all its forms forge deep connections. By contrast, global fisheries are in trouble because trawlers that roam the world have no such relationship with each other.

Some common patterns of cooperation emerge across all the examples of human and animal groups that we have considered. Cooperation can emerge in even the most unlikely situations: among enemies on the battlefield, among ferocious pirates confined at close quarters on ships at sea, and in subsistence communities where rivals compete for limited public goods. Successful cooperation depends on two fundamental rules: first, team members cooperate when it is to their individual advantage to do so; and second, every successful team requires a mechanism to suppress cheats—individuals who try to take the benefits of cooperation while avoiding the costs.

A third rule arises from the first two: cooperation is generally conditional. Soldiers will only cooperate with an enemy who reciprocates. Burying beetles only cooperate when resources allow or there is an external threat. Hadza hunters prefer to associate with people who will share. Eusocial and cooperative breeders team up only with kin. How far do these rules extend through the history of life? Every individual soldier, pirate, beetle, and bee is itself a team of cells—there are trillions of cells in each one of us. What kind of team is an individual and how do its parts cooperate?

PART II

Individuals

5

Matryoshka

In the small Russian town of Semyonov, 500 km east of Moscow, a workshop full of matrons stand at their lathes. Wood chips fly in all directions as the women swiftly and skilfully turn seasoned limewood into perfectly fashioned dolls. Over half a million of the iconic, hand-crafted Russian matryoshka dolls spill from Semyonov every year. Each larger doll contains a smaller one fitting snuggly inside it, typically making a stack of four or five dolls of ever-diminishing size. There is a myth that each stack of dolls is turned from a single piece of wood, but in reality, each doll is separately made before being assembled into a nested set. Curiously enough, Mother Nature standing at the lathe of evolution has also fashioned living organisms as nested teams of separate parts fitted inside each other, though outward appearances do not betray this.

The discovery of how life is structured as a nested hierarchy of levels and came to be this way gradually emerged over a period of 200 years, culminating as recently as 1995 in an evolutionary explanation for this structure (Figure 8). There are different ways of retelling this history, much of which has been pieced together by the Canadian historian of science Jan Sapp,[1] but however it is told what is revealed is a serpentine path littered with surprises, often folding back on itself before inching forward towards the picture we have now.

Our story begins with a Darwin. Not the Darwin that everyone has heard of, but Charles Darwin's grandfather, Erasmus Darwin (1731–1802) (Figure 9). Erasmus was a highly respected doctor, practising in and around the town of Lichfield in the English midlands. He was also well known as a poet, writing especially on scientific subjects. In his book-length poem *The Temple of Nature*, Erasmus revealed a most unorthodox idea:

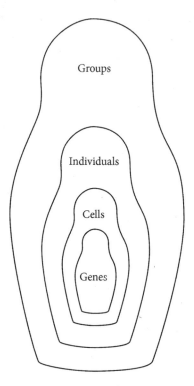

Figure 8 Life is hierarchically structured like a set of Matryoshka dolls

> Organic Life beneath the shoreless waves
> Was born and nurs'd in Ocean's pearly caves;
> First forms minute, unseen by spheric glass,
> Move on the mud, or pierce the watery mass;
> These, as successive generations bloom,
> New powers acquire, and larger limbs assume . . .[2]

In this terse and astonishingly prescient verse, written more than 60 years before his grandson Charles published *On the Origin of Species*, Erasmus suggests that complex life evolved from invisibly small, aquatic microbes.

Bacteria were first seen a century before by the pioneering Dutch microscopist Antonie van Leeuwenhoek (1632–1723), but it was Erasmus Darwin who proposed that they were the ancestors of larger, more complex life. To emphasize his point, the good Dr Darwin emblazoned the carriage that he used to make house calls with the daring motto *Ex Omnia Conchis*— 'Everything comes from seashells'. Erasmus Darwin's hypothesis was not taken seriously in his day, but within the bounds of poetic licence we now

Figure 9 Erasmus Darwin

know that he was right. Life began with microbes in the sea. A microbial cell, invisible to the naked eye, can indeed be considered the smallest and first doll in the matryoshka of life.

One might expect that grandson Charles' own research in evolution would have been inspired by the work of his grandfather Erasmus. Charles did indeed study his grandfather's work when he was a medical student in Edinburgh,[3] but in later years he was dismissive of its importance. When his book *On the Origin of Species* was published in 1859, Charles Darwin avoided his ancestor's poetic speculation on the origin of life and its hierarchical structure remained yet unrecognized. It was not till nine years later that a new discovery about something everyone ignored or took for granted opened minds to the possibility that one organism might exist inside another.

You probably walked over some today, embedded in the surface of a concrete pavement; you are likely surrounded by them growing like a crust on the exterior walls and roof of your home. Others, looking like miniature shrubs or crinkly grey leaves, live on the bark and branches of trees. Lichens are everywhere. Dissecting lichens under a microscope, the Swiss botanist Simon Schwendener (1829–1919) claimed in 1868 that they were something more special than anyone had realized. Lichens, he said, are in fact not one organism but two, comprising a fungus containing imprisoned algae.

Schwendener described the relationship in human terms, saying that the fungus was 'a parasite which is accustomed to live upon the work of others; its slaves are green algals, which it has sought out or indeed caught hold of, and forced into its service'. Other lichenologists, who had painstakingly catalogued thousands of lichen species without perceiving anything of the sort, were appalled by the idea. Schwendener had turned innocent-looking lichens into an 'unnatural union between a captive Algal damsel and a tyrant Fungal master',[4] and they were having none of it. It surely didn't help Schwendener's case that he had chosen the abhorrent metaphor of slavery when the transatlantic slave trade had only just been abolished and, in spite of very recent emancipation in the United States, millions of people remained in servitude.[5]

Despite his ill-chosen metaphor for the relationship, Schwendener was correct about the dual nature of lichens. Following the initially discouraging reception of Schwendener's discovery, other microscopists confirmed the evidence, and the dual nature of lichens was established. Ten years after the discovery, in 1878, the German Heinrich Anton de Bary coined the term *symbiosis* to describe two species that live in intimate association with one another, as in lichens. Unlike Schwendener, de Bary did not regard the alga as the slave of the lichen fungus, since he observed that the algae in lichens can be found living independently but appear to grow best when inside a lichen.[6]

It is still debated to what degree the lichen symbiosis represents a relationship of mutual benefit—a *mutualism*—or a one-sided relationship of parasitism of the alga by the fungus. It is in the nature of cooperation that there is a fine line between mutualism and parasitism. De Bary's definition of symbiosis is usefully agnostic about this, implying only that the relationship is intimate, with one partner often found living inside the other.[7]

Being particular to lichens, the discovery of the symbiosis did not directly reveal any general principle that could be applied to all species, but it did inspire biologists to interpret what they saw through their own microscopes with fresh eyes. New hypotheses followed as various objects seen inside cells were interpreted as symbiotic microbes. Maybe the cell nucleus was a bacterium? Surely the numerous tiny green objects inside plant cells (chloroplasts) must be captive algae? What about the tiny cigar-shaped bioblasts, later known as mitochondria, which so resemble bacteria? By the beginning of the new century, the idea that cells contained microbes was commonplace among biologists in continental Europe, Britain, North America and Russia, though there was much less agreement about who had first discovered what. One might think that this was the moment at which the lid finally came off the matryoshka, but in the event it was only a fleeting peek beneath her bonnet.

The moment of final realization that inside the outer matryoshka of the cell was another ought to have been when Frenchman Paul Portier wrote in 1918:

> All Living beings, all animals from Amoeba to Man, all plants from Cryptogams to Dicotyledons are constituted by an association, the 'emboitement' of two different beings.
>
> Each living cell contains in its protoplasm formations which histologists designate by the name of 'mitochondria'. These organelles are, for me, nothing other than symbiotic bacteria, which I call 'symbiotes'.[8]

In his book entitled *Les Symbiotes*, Portier claimed not only that mitochondria inside cells were bacteria but also that he could extract and cultivate them from various plant, insect, and animal cells. But whatever he was cultivating, they were not mitochondria. Although mitochondria are indeed bacterial in origin, during nearly two billion years of evolution they have become too completely integrated with the rest of the cell to replicate outside it. This was not known in Portier's time, but there was deep scepticism from other scientists, especially those who studied pathogenic bacteria in the Pasteur Institute for whom bacteria were first and foremost germs. The deepest opposition, however, came from a totally different and unexpected direction.

The year after the publication of *Les Symbiotes*, Portier's own publisher issued another book attacking it called *Les Mythes de Symbiotes*. The author was Auguste Lumière, the co-inventor with his brother Louis of the cinematograph. The success of his inventions gave Lumière deep pockets which he had used to finance the conversion of a building in the city of Lyon into his own private laboratory with a staff of 15 scientists. Lumière did a demolition job on *Les Symbiotes*, denying that mitochondria were found in all cells and ridiculing the idea that bacteria inside cells could do anything other than cause disease. The intellectual shadow of Louis Pasteur who showed that certain bacteria cause disease still loomed very large in French science, even though he had died 20 years earlier. Portier attempted to publish a new edition of *Les Symbiotes* containing a rebuttal, but the moment had passed. He moved on to study other examples of symbiosis. In a retrospective of his work published on his retirement in 1936, Paul Portier retracted the hypothesis that mitochondria were bacteria, and so flipped closed again the bonnet of the matryoshka.[9]

In 1930 the German biologist Paul Buchner published a 900-page volume detailing the myriad examples of symbiosis between various insects and the different microbes inside them, without which they could not survive on deficient and unvaried diets of wood, plant sap, or blood.[10] But in the ensuing 30 years, the hypothesis that mitochondria and chloroplasts found inside cells were bacterial and symbiotic in origin was all but forgotten. In the words of a leading textbook issued in 1925, 'to many, no doubt, such speculations may appear too fantastic for present mention in polite biological society; nevertheless, it is in the range of possibility that they may someday call for more serious consideration'.[11] That day was indeed to come, and these prophetic words became a clarion call, but not before a revolution in genetics that changed everything.

The revolution hinged on the famous discovery in 1953 of the double helical structure of DNA.[12] Jim Watson and Francis Crick ended their brief report on the double helix with the gnomic statement that '[i]t has not escaped our notice that the specific pairing we have postulated immediately suggests a possible copying mechanism for the genetic material'. The proposed mechanism was revealed in a second report published a month later that showed how the genetic code is embedded in the structure of DNA.[13] Nature's genetic secrets are encoded in the sequence of molecules strung along the DNA chain like the letters strung together in this sentence. Particular sequences of letters in the code represent genetic instructions that specify how cells are made and operate.

The DNA sequence differs from a sentence of text in that there are two copies of each DNA sentence, one encoded on each strand of the double helix. The equivalent in a book would be like having each line of text spelled out twice, one copy being a mirror image sitting alongside the other. If you opened such a book and found that it was written in a language you could not understand, you would still be able to see that the lines had this double structure, even if you could not actually decipher the text. This is what Watson and Crick found. The structure of DNA and of the genetic code was revealed, but it was 20 years before chemists were able to devise methods to read the coded sequences that the structure contained. Nonetheless, Watson and Crick's discovery was revelatory enough. It was now completely clear where the genes responsible for all the essentials of life would be found: in DNA. This was soon to produce the strongest evidence yet for the symbiotic origin of mitochondria and chloroplasts.

Spirogyra sounds like the name of a giddy fairground ride, but it's a freshwater alga called 'mermaid's tresses' by its admirers and 'blanket weed' by the owners of garden ponds that have become overrun by the slimy stuff. The tresses are long hairlike chains of algal cells that under a microscope can be seen to contain a spiral ribbon studded with bright green chloroplasts. The gyrations of a spiral nicely describe the twists and turns of history by which, in the end, the hypothesis that chloroplasts derive from symbiotic bacteria finally came out on top. Fittingly, it was in this plant that it was discovered in 1959 that chloroplasts contain their own DNA.[14] This meant that there were probably genes inside chloroplasts, but why on Earth were they there?

The conclusion was that these were the legacy of a bacterial origin. From that point on, more and more evidence accumulated of the similarity between chloroplasts and a group of photosynthetic microbes called cyanobacteria. Chloroplasts even multiply themselves within plant cells. Not everyone was convinced, though. A couple of sceptics complained that the theory had been circulating like a bad penny for a long time. The other side of this bad penny was of course the idea that mitochondria were also originally bacterial symbionts. Soon after the discovery in chloroplasts, DNA was found in mitochondria, too.

With the benefit of hindsight, it might seem as though the new discoveries of DNA in chloroplasts and mitochondria should immediately clinch the theories of their symbiotic origin. However, most scientists are conservative in their ideas and always look for alternative explanations that could reconcile new facts with existing theories. Cynics say that science advances one funeral at a time. Unconventional ideas need a champion, and in this case her name was Lynn Margulis.

Lynn Margulis (1938–2011) (Figure 10) was a masters student at the University of Wisconsin when she heard the challenging words from the 1925 textbook that ideas once too fantastic to mention in polite society 'may some day call for more serious consideration'. Margulis later said that on hearing this, 'the course of my professional life was set forever!'[15] She dedicated her life to the idea that symbiosis with microbes is the main source of evolutionary novelty—that it is not just responsible for chloroplasts

Figure 10 Lynn Margulis

and mitochondria but much else besides. In 1967, after 15 rejections by various scientific journals, Margulis at last published her work. It contained the most complete theory yet, based on evidence from Earth history as well as cell biology, on how cells acquired mitochondria and chloroplasts.[16]

Publication of Margulis' paper represented the moment when the matryoshka was finally and irreversibly opened for all to see inside. It revealed the truth of Paul Portier's claim, made and then recanted a generation before, that all animals and plants 'are constituted by an association, the "emboitement" [interlocking] of two different beings'. We can now be much more specific than Portier could possibly have dreamed about how this happened.

For the first couple of billion years of life on Earth, simple cells called prokaryotes were the only kind of individual in existence. Prokaryotes include bacteria, which are still by far the most numerous organisms on the planet. They have rich social lives as we shall discover later, but they are essentially singletons who like to party. Then, about 1.8 billion years ago, two different kinds of prokaryote teamed up and became permanently hitched in a new kind of cell: a eukaryote. Eukaryotes are sexual and their cells complex. Sigmund Freud (himself a eukaryote) would have loved them. Inside eukaryote cells, the genes are protected within a membranous bag called the nucleus, and the cytoplasm around the nucleus contains a skeleton made of scaffolding tubes that can be reconfigured to move things around or change the shape of the cell to engulf prey. The superpower that makes all this possible is the energy provided by mitochondria.

Then, over half a billion years later, a eukaryote cell acquired a second inhabitant—a cyanobacterium with a talent for photosynthesis.[17] The cyanobacterium evolved into the chloroplast and its host became the earliest ancestor of green algae from which land plants later evolved. Today, the chloroplasts of marine algae and land plants between them remove more than 100 billion tonnes of carbon from the atmosphere every year.[18] If you consider that all macroscopic life is composed of eukaryote cells and that all animals feed directly or indirectly on plants, it would be hard to disagree with Lynn Margulis that symbiosis is of monumental evolutionary importance.

You will have noticed that we have been bandying with billions. Prokaryotes ruled unchallenged for two billion years. Eukaryotes had appeared by 1.8 billion years ago. It was half a billion years more before chloroplasts were formed. These were rare evolutionary events. Symbiosis is common—we shall look at many examples in later chapters—but what is uncommon in these two particular cases is that each permanently formed a new kind of individual. The first made eukaryotes out of two prokaryotes. The second made plants from a eukaryote and a prokaryote. Before each of these events, the symbionts led separate lives, afterwards they were bound together and could only reproduce as a team. That team was the new individual—a new matryoshka added to the family.

In 1995, John Maynard Smith and the Hungarian evolutionary biologist Eörs Szathmáry proposed that the formation of a new kind of individual represents a major transition in evolution (MTE).[19] Once established, MTEs tend to be hard to reverse because the members of the team often evolve dependence on each other. Each doll becomes enclosed by the next in a chamber with no exit.

This hypothesis, which like many of the most elegant ideas seems obvious once you have heard it, can explain why life is constructed like a nested hierarchy of matryoshka and how cooperation and team formation bring it about. There have been refinements to Maynard Smith and Szathmáry's original idea in the 30 years since they proposed it, such that there is now a consensus about the recognized MTEs.[20] These include when genes teamed up to make chromosomes; the two events that first produced eukaryotes and then made some of them plants; the evolution of organisms made up of many cell types—multicellularity—which has occurred several times independently among single-celled eukaryotes; the formation of social groups in eusocial ants and bees and arguably even our own species; and last but certainly not least, lichens, corals, and many other symbioses. These MTEs give us a rich seam of examples that we mine in the following chapters to explore how major transitions in evolution happen and why they are important.

Here is foretaste of what to look out for in the coming chapters. There are two distinct and by now familiar steps in major transitions. The first step is always the formation of a team within which cooperation is favoured by a division of labour among teammates.[21] In some teams, such as the bees in a nest or the cells in our own bodies, all members are kin (closely related) or clones (genetically identical). In others, there is a symbiosis between quite unrelated species such as bacteria and eukaryotes—what might be called the odd couples.

Symbioses between odd couples of unlike partners are ubiquitous. That should be no surprise, given that team formation requires team members to show a division of labour, so providing complementary resources. This division of labour is likely to be strongest between partners that are most different from each other to begin with, such as algae and fungi. But there is a problem in explaining the evolution of symbioses. Within a species, altruism may evolve through kin selection, but how can altruism evolve between totally unrelated organisms such as algae and fungi?

Darwin referred to this problem and wrote that if some property belonging to one species could be shown to exist solely for the benefit of another, his theory of natural selection would be disproved. One obvious answer is that apparent altruism between species is actually no such thing. A symbiosis that provides mutual benefits—'I'll scratch your back if you scratch mine'—is a transaction, not an altruistic proposition. Alternatively, one-sided benefits may represent parasitism in which the loser is robbed. 'I'll ride your back' is not an altruistic proposition either. Is either of these explanations correct? And how do symbioses evolve? Can parasites become symbionts, or the reverse?

The second step in a major transition is the transformation of the team into a new entity, which requires the suppression of conflict among team members. Suppression of conflict involves controlling the ability of team members to branch out on their own and to reproduce independently. This is how teams become a new kind of individual. Suppression of conflict happens in two distinct ways, depending on whether the different members of the team are kin or clones on the one hand or unrelated on the other. In the former case, the power of kin selection can ensure cooperation, but in the second case something else is needed to tame the inevitable differences in an odd couple. What is it?

6

Odd couples

They are parasites and rotters, the lot of them! This is a scientific description of fungi, not an indictment. Parasitic fungi obtain their sustenance
from the living bodies of other organisms, while the rotters, or more technically 'decomposers', feed on the dead. Some do both. The honey fungus
Armillaria mellea parasitizes living trees and then feeds on the decomposing
remains of its victims after it has killed them. On the face of it, fungi may
seem like the most unpromising candidates to join a team of any kind.
Fungi have form and have even been captured parasitizing other fungi in
400-million-year-old fossils. Who would expect cooperation from species
that, when they are not pillaging, are feasting on corpses? Team Zombie is
no one's voluntary choice. But deeper consideration of how fungi, especially the parasitic ones, make their living reveals that on the contrary, fungi
have some unique life skills that make them potentially valuable team
members.

Fungi are good at getting up close and personal with other organisms. All
organisms have defences against the natural enemies that want to call them
lunch, so parasites need to evolve ways around the defences of their victims
or 'hosts'. Once a parasite has achieved this, there is then selection on the
host to raise its defences further—potentially leading to an arms race
between them. A consequence is that parasites are channelled by natural
selection towards greater and greater specialization in their choice of host.
Most parasites of all kinds, not just fungi, can only attack a small range of
hosts, sometimes as specific as just the small fraction of a host population
that lacks one particular gene for defence.

Lichens were the first symbiosis to be recognized as a team of two unlike
organisms, forming a unit with an individuality all of its own. How did
fungi team up with algae to make lichens? Recall how the Swiss botanist
Simon Schwendener was pilloried for describing lichens as master–slave

relationships in which the algal partner is the captive of the fungus. Whether that is an accurate description or not, it does seem quite likely that in the formative stages of its evolution, the lichen symbiosis may have begun with the fungus parasitizing the alga. This hypothesis is supported by the fact that the fungal partners are specialized 'lichenized' fungi belonging to many thousands of different species, while the algal partners belong to a relatively small number of unspecialized species that are often found growing on their own. Extreme specialization is typical of parasites.

It seems to be inevitable that in trying to understand what is going on inside a lichen, we reach for analogies for the relationship such as master–slave, but such comparisons can be misleading. Lichens of one kind or another are to be found everywhere, especially in places where nothing else can grow, such as on the most exposed rocks on the seashore that are pummelled daily by the ocean, or inside the very surface of stones where they can somehow make a living. It has been argued that this demonstrates benefit to the alga because it could not grow in such places on its own. That is the essence of symbiosis and renders analogies redundant.

The transformation stage of a major transition is complete when both partners become so dependent on each other that they reproduce as a unit. Typically, the smaller partner becomes a passenger transmitted from one generation of its host to the next along with the genes of the host, as happens in mitochondria and chloroplasts. What happens in lichens is complicated and different lichens seem to be at different points on the road to a fully transformed, major transition.

The lichen symbiosis has evolved many times over the last 250 million years, many events producing a major transition when the union between what were once two independent organisms became one. There are reckoned to be at least 27,000 lichen species, mostly evolved in the last 60 million years. The number of lichens is similar to the estimated number of ant species, the second most successful group among insects after the beetles. Lichen diversity is the result of fungal rather than algal evolution because the different fungal species involved outnumber their photosynthetic partners, known as photobionts, by a ratio of 10:1. The same photobionts show up again and again with different fungal partners or 'mycobionts', making different lichens. Such new symbioses can spawn hundreds of new lichen species as the fungus evolves to fill new niches. What was until recently believed to be a single species of lichen has been revealed by genome sequencing of the mycobiont to probably represent more than 400 (Plate 6).[1]

The great majority of lichenized fungi belong to the cup fungi (ascomycetes) and the rest to the gill fungi (basidiomycetes). Likewise, the photobionts belong to two different groups: most are algae (eukaryotes) and a smaller number are photosynthetic cyanobacteria like the ancestors of chloroplasts. The lichenized fungi are not found living on their own without a photobiont, but conversely most of the photobiont algae can be found living independently as well as in lichens. The photobiont cyanobacteria also have a foot in both camps. Most cyanobacteria found as photobionts can also be found free-living, but those in the genus *Rhizonema* are only found in lichens.[2] Dependent though they are on lichens, *Rhizonema* are not that particular about which mycobiont they team up with—their evolutionary history reveals that they have often switched between different lichens.

How many matryoshka dolls are there in a lichen? In the majority in which the photobiont is an alga, the answer is at least five: the mycobiont and its mitochondria, plus the photobiont, its mitochondria, and chloroplasts. In fact, the count for the entire team could be seven or more, because it has recently been established that many lichens contain not just the mycobiont but yeast (a single-celled fungus) as well which feeds on the photobiont.

If your head has begun to swim with the complications of lichen relationships, then I have some more bad news when it comes to lichen romance. In her sonnet 'How Do I Love Thee? Let Me Count the Ways', Elizabeth Barrett Browning managed to tot up a mere seven modes of devotion to her husband Robert Browning. A recent count of the ways in which lichens reproduce found 27. Of course, lichens have the big advantage of numbers over one poet. This is not a competition, but it is a measure of how miscellaneous lichens are. Of the 27 modes of reproduction found in lichens, 13 involve a dispersal unit that contains the cells of the photobiont encased in fungus.[3] In the remaining 14, the unit is a lone fungus that must find free-living cells of the photobiont to make a new lichen. Let me round out this conceit by observing that coincidentally there are also 14 lines in a classical sonnet such as Barrett Browning's.

As far as major transitions go, the 27 modes of lichen reproduction can be divided into just two kinds, depending on whether the photobiont is transmitted vertically (by inheritance) or horizontally (acquired from the environment). The lichens in which the photobiont is vertically transmitted with the mycobiont have completed the transformation to a new individual.

The lichens in which the photobiont is horizontally transmitted, and so must be recruited anew in each generation of the mycobiont, are on the road but not there yet.

Are there photobionts that resist lichenization or get the upper hand over the mycobiont? We simply don't know, but there are fungi that have given up the lichen habit during their evolution, so there must be costs to the symbiosis that favour terminating it in some circumstances. For example, in an environment such as the soil, which is too dark for the photobiont to earn its keep and where alternative food is available for the fungus.

Many symbiotic relationships are intimate, and yet like the photobionts in many lichens, at least one of the partners is recruited from a free-living population. In the seas around Hawaii lives a small, mouse-sized squid called the bobtail (Plate 7). This little creature spends the daylight hours in a camouflage jacket, buried in the sand, only emerging into the water column at night to feed on shrimp. Not a bad life, but the Hawaiian bobtail squid has many enemies, even under the cloak of darkness. It is prey to fish patrolling the waters at night that can pick off from beneath any small animals that are silhouetted against the moonlit sky. This is dangerous for bobtails, but the squid have an ally in their defence: bioluminescent bacteria. You might think that the last thing that an animal trying to hide at night needs is the services of bacteria that emit an eerie glow, but here is the surprise. The bacteria live inside a specialized light organ inside the squid, from where light can penetrate its near-transparent body, illuminating the underside of the animal at night. This disrupts the smooth, alluring silhouette of the squid, hiding it from preying eyes.[4]

Bobtails are not born with a resident infection of bioluminescent bacteria but acquire the microbes when they are still only 2 mm long hatchlings. The sea is full of thousands of species of bacteria, but of all the bacteria in all this underwater world, it is only bioluminescent *Vibrio fischeri* that colonize the hatchling squid. This is no coincidence. The tissues of the squid are able to recognize *Vibrio fischeri* cells, trapping them in mucus laced with antibiotics that eliminate unwanted species of bacteria. From the exterior, the selected bacteria migrate through six pores in the skin of the squid to embark on a journey along narrow ducts reminiscent of the subterranean journey in *Indiana Jones and the Temple of Doom*. More antibiotics protect the ducts from unwanted microbes, while hairlike cilia aid the selected bacteria as they swim along. At the end of the first duct, bacteria must squeeze through another pore before entering a further duct that opens into a broad antechamber. The Lost Ark is near, but not all the travellers can reach it. Each antechamber narrows into a bottleneck, forcing bacterial cells to progress in single file towards a crypt that will only admit one cell. After the solitary cell has entered, the crypt is sealed by the bottleneck tightening behind it and the bacterium loses its tail-like flagellum. It is now trapped, unable to swim. Is this any way to treat a friendly symbiont? It is if you want to select only the right bacteria for the job.

The processing of a soup of billions of diverse bacteria in seawater down to just six cells of one bacterium, *Vibrio fischeri*, is a remarkable example of partner selection, but what happens next is even more extraordinary. The arrival of the symbiont in the squid's light organ stimulates changes to accommodate the bacteria. They are housed and fed, and multiply rapidly, swelling each crypt with innumerable copies of the founding cell that made it through the obstacle course. Twelve hours after the crypts were first colonized, the bacteria that now number in tens of thousands switch on their lights in unison.

How do thousands of bacteria imprisoned in six different crypts synchronize a light show? Bobtail squid spend the day hidden in the sand, so the cue that triggers its symbionts to switch on their lights could be the emergence of the animals into the water at dusk, but the truth is less obvious and much more interesting. More than 50 years ago, it was discovered that when these bacteria are grown without squid in a laboratory flask, they only begin to bioluminesce after some hours of culture, when the density of cells has grown a thousand-fold or more.[5] This phenomenon, first discovered in *Vibrio fischeri*, is called quorum sensing (QS), and we now know that it occurs in some form or other in many bacteria, not just bioluminescent ones.

The significance of QS is that it is a social phenomenon used by bacteria to signal to each other that they are surrounded by many kin. We explore later how QS enables cooperation to evolve among bacteria, but for now the relevant point is simply that the on-switch for the bobtail squid's lights is under the direct control of its bacterial symbionts themselves, not the squid. Imagine what might happen if the lights in your home were controlled by your pet dog and you can see why this arrangement might not be ideal. But evolution does not make ideal arrangements, it just makes do.

Given that the pet has control of the lights, so to speak, how does the bobtail squid turn them off at dawn? It reopens the crypts in its light organ and expels the bacteria into the sea. The morning after the night before,

95 per cent of the bacterial cells are unceremoniously shown the back door. The remaining 5 per cent of bacteria then have 12 hours to multiply and repopulate the light organ in time for the resumption of night-time illuminations.

There are two layers of cooperation in the squid–*Vibrio* symbiosis: first, between the bacteria which synchronize their bioluminescence via QS; and second, between the squid and bacteria, which trade light for food. So far as it can be, the squid is in control of both of these. The ordeal through which the bobtail squid force their symbiont bacteria when they are first gathered from the sea ensures that each crypt is filled with clones from a single founding cell, so they are those closest of kin. This makes it harder for an uncooperative mutant to appear and ensures that bacteria cooperate with each other to the advantage of the host. This is not the only control the squid has, though. The organ that houses the bacteria is sensitive to light, using the same light-sensing molecule found in the retina of the eye. This gives the squid the ability to monitor the performance of its symbionts, and if their light begins to dim with time, the squid can expel them and restock.

Intricately evolved though it is, the squid–*Vibrio* symbiosis has not reached the point of transformation into an MTE because the partnership does not persist throughout the life cycles of the partners. For the squid, the relationship is renewed with each generation of newborns and most of the bacteria are expelled after only 24 hours. It has been conjectured that by raising and expelling symbiotic bacteria each day, squid might be seeding the ocean with *Vibrio* bacteria that have been selected to be especially cooperative symbionts. If that does happen, this cooperation is not directed at bobtail squid in particular because the relationship is not an exclusive one. *Vibrio* also teams up with other species of squid and over evolutionary time *Vibrio* has bed-hopped between hosts.[6] In evolutionary terms, this symbiosis is more of a serial one-night stand than a committed relationship.

The squid–*Vibrio* symbiosis illustrates the two cornerstones of cooperation: mutual interest and the control of cheating. It also has a feature that seems to be important to successful cooperation between hosts and microbes: a compartment where the host can contain and control microbial activity.[7] Microbes are arch opportunists and can thrive in some surprising places. A French scientist studying bioluminescent bacteria in fish even became infected with the bacteria himself and it is said that he passed glowing stools for months.[8] He eventually rid himself of the infection but at the cost of a shining reputation.

Bobtail squid–*Vibrio* have formed cooperative bonds in an intricately evolved relationship, but this has not transformed into an MTE. Why not? For teams of odd couples to transform, potential conflict between symbiotic partners must be controlled and the two independent lineages need to be locked together. This has not happened in *Vibrio* and squid or even, despite appearances, between the mycobiont and the photobiont in most lichens. It is, though, very common in insects.

Insects are airliners for microbes, which travel in the gut, and just like an airliner, parts of the vessel are more hospitable to passengers than others.[9] Some of the passengers are stowaways, such as the bacterium *Xylella fastidiosa* that hitches a ride up front in leafhoppers, infecting the plants that the insect feeds on. *Xylella* is a pathogen capable of infecting hundreds of woody species and has devastated olive trees in southern Italy. In the midsection of the gut is the galley, which is full of enzymes that digest the insect's food. This is a hostile environment for most microbes, especially in insects such as fruit flies that feed on bacteria. The hindgut is economy class, where microbes are crammed at maximum capacity, and many are working their passage. This is where termites and other wood-feeding insects carry the microbes that digest wood fibres and cellulose (Figure 11), something that no animal can do for itself.

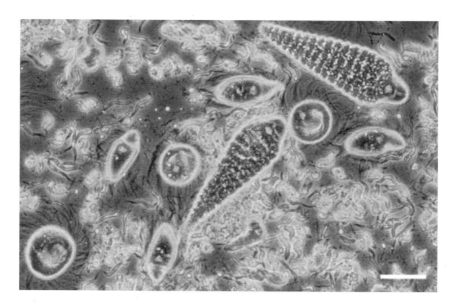

Figure 11 Symbionts in the termite gut

Termites resemble ants in appearance and like ants they are eusocial, with queens, workers, and soldiers inhabiting large nests, some of which have built-in air conditioning. But termites are actually cockroaches, not ants, and the resemblance between the two is a classic case of convergent evolution, where natural selection has led quite different lineages down the same path of adaptation. The evolution of eusociality in the termites, which of course constituted a major transition all of its own, caused the cellulose-digesting microbes to be transmitted generation to generation with their hosts, thus binding the symbionts to one another in mutual dependence. The microbes in question are protists (single-celled eukaryotes) that are lost from the termite gut each time the insect moults. Newly moulted termites reacquire the protists essential to their nutrition by feeding from the hind-gut fluids of a nestmate. This kind of vertical transmission of a symbiont between kin would not have been possible before the evolution of eusociality.[10] Thus, one MTE seems to have prepared the way for another.

The most intimate and dependent bacterial symbionts in the insect airliner have their own special cabin—an organ called the bacteriome, which is located in the abdomen of the insect. Here, bacteria are individually swaddled in host membranes, pampered and fed like first-class passengers. But the location of the bacteriome in the insect undercarriage is a tell-tale sign of their true status: this is not so much first class as a cargo hold and the bacteria are as much captives as passengers. In fact, the bacteria are transmitted generation to generation through the eggs of their hosts, which is the ultimate sign of a symbiosis that has transformed into an MTE.

What service do the symbionts in the bacteriome render their hosts? There is a clue in the genome of the bacteria, such as that of *Buchnera aphidicola* found in aphids. *Buchnera's* genome is very small compared to those of free-living bacteria. In the 150 million years since the bacterium teamed up with aphids, it has been stripped of many genes needed for independent existence, but it does still contain the genes needed to manufacture essential amino acids. Insects, like many other animals including ourselves, are unable to manufacture half of the 20 amino acids that are required to make proteins. There are only two sources of these missing essentials: food or symbionts.

Aphids, leaf hoppers, and other sap-feeding insects sup from a fire hose that pumps them full of a dilute sugar solution containing little else. The excess solution spills from the rear of an aphid like beer from a spigot. Aphids have a lollipop diet—all sugar and no essential amino acids. However, thanks to their bacterial symbionts, aphids can live happily on sugar. The proof that their symbionts fulfil this role is that if experimentally dosed with antibiotics that kill *Buchnera*, aphids need dietary supplements to survive.[11]

Symbiotic microbes that supply nutrients are found in most, if not all insects.[12] Part of the reason for this may be that being small, insects often specialize on a very narrow range of foodstuffs that cannot supply all the essentials. Even human blood, which you might think ought to contain everything a parasite could wish for and then some, is deficient in B vitamins. Bed bugs, lice, and tsetse flies all need microbial supplies to supplement their diets.

As ever, where there is cooperation there is the risk that one or other partner will run amok, and this is certainly an ever-present danger from bacteria, which can multiply and evolve at a rate that no multicellular organism can remotely match. Perhaps because symbioses between microbes and multicellular organisms are so inherently unmatched in this respect, they are also unbalanced in another: who is in charge? It seems an invariable rule that in odd couples, the larger symbiont controls the smaller one. We saw this in the way that bobtail squid filter, control, monitor, and eject *Vibrio* bacteria to suit the animal's needs.

Cultivating bacteria inside your body is like playing with fire: if you don't control the fire, it will either go out or burn down the house. The maize weevil *Sitophilus zeamais* infests cereal grains, particularly maize kernels, and is a serious pest all over the world. It harbours bacterial symbionts in a bacteriome where their spread is controlled by the insect's immune system. Experimentally disabling this immunity causes the bacteria to proliferate, break out of the bacteriome, and invade the body of the weevil, burning down the house.[13]

There is a continuum of benefit in symbiosis, from cooperation (mutualism) at one end to theft (parasitism) at the other. It's hard to tell where the microbial symbiosis found in the bacteriome of insects sits on this continuum because the bacteria are so completely controlled by the insect host. But experiments such as the one with the maize weevil suggest that, if allowed to, bacteria would multiply much faster than they can in the bacteriome and that therefore they are the prisoners not mutualists of insects.[14]

How do symbioses such as the ones seen in the bacteriome evolve in the first place? Do the bacteria start out as parasites, or are they independent, free-living organisms such as the photobionts recruited from the environment by lichen fungi? The evolutionary history of symbiotic bacteria can be reconstructed from their genome sequences, and it gives a clear answer to this question. Most symbiotic bacteria evolved from parasitic ancestors. The transition from parasite to symbiont has happened at ten times the rate at which free-living bacteria have given up their liberty.[15] Thieving bacteria often end up in symbiotic jail, working a treadmill for the benefit of their erstwhile victims. We mustn't be too taken with this analogy though because it takes thousands of generations for evolution to serve justice on the thieves.

It may seem odd that a parasite can be co-opted for the benefit of its host, but the reason that this has so often occurred is that parasites and hosts begin with a strong interest in common: the survival of the host. If the host dies, that is no good for the parasite. For this simple reason, any adaptations that the parasite can evolve that prolong the life of its host are to its benefit too. When autumn paints the leaves of apple trees in yellow and red, the days are numbered for a tiny moth caterpillar that lives inside them. However, the larvae have a trick up their ravelled sleeves. Each one maintains a small island of green around itself that provides enough food for the larva to continue feeding until it can pupate. This salvation is not the work of the larva itself but of its bacterial endosymbiont that manipulates the plant hormone controlling leaf senescence. When moth larvae are rid of their bacteria with an experimental dose of antibiotic, no green island forms and larvae run out of food and die.[16] With shared interests such as this between larva and bacterium begins the alignment of interest between host and parasite that can ultimately lead to cooperation.[17]

As a parasite begins to evolve beneficial qualities for its host, it starts a journey along the symbiotic continuum towards mutualism. The early steps of the process by which parasites become beneficial have actually been observed in action. In 1972, Kwang Jeon at the University of Tennessee reported that he had noticed some years earlier that a lab culture of amoeba—single-celled eukaryotes that feed on bacteria—had become infected by rod-shaped bacterial parasites later found to be a species of *Legionella*.[18] When first observed, the *Legionella* infection slowed down the rate at which the amoebae reproduced, demonstrating that the bacteria were indeed parasitic. Five years later this colony, still heavily infected by *Legionella*, was back to its previous rate of growth. It looked like the parasites had become benign. Jeon extracted the nucleus from infected amoeba cells and transplanted them into uninfected cells from which the normal nucleus had been removed. The resulting hybrid amoebae grew poorly, but when these cells were injected with bacteria they behaved normally. The conclusion was that bacteria and amoebae had become adapted to one another, now forming a mutualism in which each required something from the other.

A similar evolutionary change has been observed in wild populations of fruit flies following the introduction of a bacterium called *Wolbachia pipientis*. *Wolbachia* is an intracellular parasite found in a huge range of insects. It is the same bacterium that prevents chlorophyll loss around leaf mines in senescing leaves. Its success seems to be due to it being invisible to the insect immune system, plus an ability to survive outside the insect body, where it can infect other insects, woodlice, and even nematode worms. The bacterium gets into the eggs of its host and, once present, it is passed down the generations along the female line.

Populations of fruit flies in California were free of *Wolbachia* infection until the 1980s when it started to spread. When it first appeared, *Wolbachia* reduced the number of eggs laid by infected females compared to uninfected ones. Twenty years later, the difference had been reversed and infected females laid 10 per cent more eggs than uninfected ones.[19] So here too, a new symbiosis formed with a parasite and then rapidly progressed towards a mutualism with its host. These events happened fast enough to be observed by scientists because the hosts in both cases—amoeba and fruit flies—have short life cycles, speeding up the rate at which they can evolve.

The *Legionella* bacteria in amoebas and *Wolbachia* in insects live inside the cells of their hosts, so they are referred to as endosymbionts. Due to their presence inside cells, endosymbionts are able to infect the eggs of their host, with profound consequences. Vertical transmission from generation to generation binds host and symbiont into a team that reproduces as a unit—producing an MTE.

The formation of the team aligns the evolution of symbiont and host, but their interests are not identical because there is one very significant difference between them. The genes of the host, if it reproduces sexually, are transmitted through both males and females: sperm and eggs. However, the symbiont is only transmitted via eggs. Every new individual in the host population comes from an egg, so the symbiont can spread into both males and females by this route, but once in a male it reaches a dead end because males don't make eggs. So here we have a conflicted team in which the members fight for their own advantage. *Wolbachia* aims its punches below the belt.

Wolbachia manipulates the reproduction of its insect hosts in a variety of ways that increase its transmission.[20] In some insects, *Wolbachia* kills male embryos, giving an advantage to female larvae by terminating with extreme prejudice the brothers with whom they would otherwise have to compete for food. In other insects, *Wolbachia* alters hormones to feminize genetic males so that they lay eggs that will carry the bacterium. In insects such as bees and ants where males are haploid (one set of chromosomes) and females diploid (two sets), *Wolbachia* turns males into females by doubling their chromosomes. Some infected wasp populations have become all-female as a result but can be returned to normal by experimentally applying an antibiotic that kills *Wolbachia*. In other such cases, applying antibiotic simply renders the wasps infertile because they have become so dependent on *Wolbachia* for reproduction that they have gone beyond the point of no return.

If harbouring endosymbionts can be a mixed blessing for a host, it is hazardous for the endosymbiont too, particularly if it is only capable of vertical transmission. Over time, endosymbionts that formerly lived independent lives tend to lose genes that they no longer use. Parasites become mutualists, mutualists become dependent, and their redundant genes rot under the pressure of mutation. The rump of the bacterial genome containing only the small number of genes essential to the maintenance of the symbiosis may survive for tens or even hundreds of millions of years.[21] But eventually even important genes may be lost: transferred to the host's own genome where they function better for the host. Endosymbionts themselves can be lost in this manner, leaving nothing but the grin of a Cheshire cat.

As the Cheshire cat in Lewis Carrol's *Alice in Wonderland* fades from view, leaving only her grin upon the branch where she sat, another tree-dweller appears, drawing our attention with his incessantly shrill chirrup. This creature is a cicada, loudly advertising his presence to females who have just emerged from the soil beneath, where, as juveniles, they have until now feasted and grown fat on the sap of tree roots. Cicadas have inspired poets for millennia, from ancient China to ancient Greece, but the cicada story we are interested in is much, much older.

Once upon a time, 260 million years ago, the ancestors of modern cicadas and all their tribe of sap-sucking bugs filled the bacteriomes in their bellies with bacteria. Over time, in cicadas, the genomes of these bacteria shared the universal fate of endosymbionts, shrinking to the edge of existence. Some were then tipped over the edge by the appearance of a new kid on the block. There are always other parasites around, ready to step into the shoes of an ageing endosymbiont. Among cicadas, which spend the majority of their lives in the soil, parasitic fungi are a ready-at-hand source of infection. In multiple cases, such fungi have replaced bacteria to become endosymbionts of cicadas. In some species of cicada, the process has even repeated itself, with endosymbiotic fungi replacing each other.[22] Fungi, as we shall see, get in everywhere.

7

Phytosympathies

The thousands of sightings of UFOs reported in the United States this century come mainly from large cities.[1] Aliens arriving from the sky love the bright city lights, or perhaps they *are* the bright city lights. Not so the aliens of the Devonian age that appeared more than 410 million years ago. They loved the light alright, but their fossil remains are lodged in rural field walls and rocks in the tiny Scottish village of Rhynie. They were some of the first vascular plants. The pioneers were aliens on *terra firma*: little green beings, quite unlike anything seen on land before—no less alien to land than a fish out of water or a Martian from outer space. Arriving from the sea rather than from the sky, these aliens did not travel alone; they brought symbiotic fungi with them. On the same principle that symbiosis makes lichens such superlative pioneers, the fungal symbionts of the first land plants were probably crucial to their success in a wholly new environment.

Fossils of the early land plants at Rhynie are found in a fine-grained, silicate rock called chert that was laid down in a hot-water spring. The remains of plants and small animals living in cooler water around the spring were preserved in extraordinarily fine detail as their cells were gradually infiltrated and replaced by silica. The plants appear to have had primitive roots, but they mostly lacked leaves or had only small leaf-like outgrowths of the stem and were rarely more than 50 cm tall (Figure 12).[2] Examine sections of the stems under a microscope and hyphal threads of fungi are revealed, penetrating the tissue and branching into tiny bush-like growths inside cells, just as they do in the roots of modern plants. The ancient fungi found in the early land plants from Rhynie are the very image of the modern-day symbionts known as arbuscular mycorrhizal (AM) fungi. Over 70 per cent of land plants still have AM fungi today.

The AM fungi of modern plants supply their hosts with nutrients gathered from the soil, particularly phosphorus, and in exchange receive carbohydrates

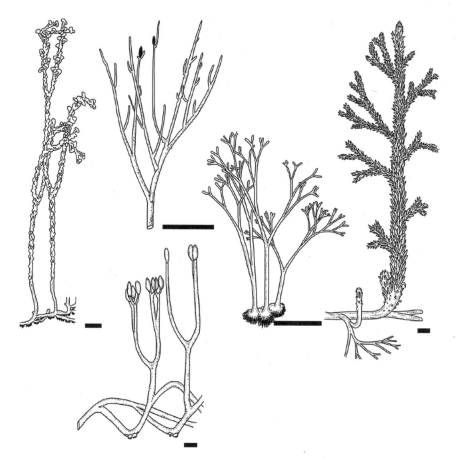

Figure 12 Fossil plants of the Rhynie chert (scale bars = 10mm)

and other compounds that the plant manufactures through photosynthesis. As much as 20 per cent of a plant's production can go towards feeding its AM fungus, but it can receive back up to 80 per cent of the nitrogen and all the phosphorus it needs.[3] AM fungi also make their hosts more resistant to drought and pathogens.[4] There is no reason to doubt that there was a basket of mycorrhizal benefits in Rhynie 410 million years ago, too. AM fungi all belong to a single, very ancient lineage that is even older than the land plants seen in the Rhynie chert. These fungi are obligate symbionts, unable to complete their life cycle without their host plants. The association, stretching way back, possibly to the marine ancestors of land plants, could be 500 million years old or more.[5] Since all life on land ultimately depends upon plants, it is hard to overstate the importance of the AM fungal symbiosis; but was its evolution a major transition, reproducing as a single entity?

Although plant and fungus are nutritionally dependent on each other, they do not reproduce as a unit. This means that, important as it is, the formation of the symbiosis is not an MTE in which a team becomes so tightly bound that it becomes a new kind of individual, as has happened in some lichens. Inside plants, AM fungi are confined to roots and do not find their way into the seeds or fern spores by which their hosts reproduce.

Until recently it was thought that AM fungi were asexual, but it now appears that sex does happen at least occasionally, hidden away inside roots.[6] Fungal spores appear at the tips of infected roots, packed full of thousands of fungal nuclei and often carrying a passenger, too. The passenger is a bacterium with the hideous Latin name *Candidatus Glomeribacter gigasporarum* that it is kinder to all concerned to call by its initials *CaGg*.[7] The bacterium is a symbiont of AM fungi that assists fungal spores to find new plant hosts but, for reasons unknown, only some AM fungi carry it. One must suspect that, like other symbionts, it carries a cost and that this sometimes outweighs any benefit.

Some AM fungi appear to reproduce only by spores, but many can infect new hosts from the existing mycelium in the soil or from infected root fragments. Once liberated into the soil, the AM fungal spore germinates and sends out an exploratory mycelium. If the spore is carrying *CaGg*, this seems to aid the growth of the mycelium. Most species of AM fungi do not seem to be fussy about the species of host they team up with. If they encounter a root, helped by a chemical signal issued by the host, the mycelium enters the root and from there proliferates in the soil around it. The threads (hyphae) of the fungal mycelium often fuse to each other where they touch, forming a network in the soil that connects plants to each other. Individual plants may harbour many different AM fungal species in their root system and the same fungi may infect different host species. The resulting web of connections forms a common mycorrhizal network that sounds like a cooperative fiesta where everyone is linking arms, singing 'Auld Lang Syne' and sharing a potluck supper, but it's not quite that simple.

Although the common mycorrhizal network creates connections between different species of plant and fungus, there is no common pool of resources where everyone is supping from a common pot through a straw. That could produce a tragedy-of-the-commons situation in which cooperation can be totally undermined by selfishness. Rather, each fungus and each plant has control over what it exchanges with any individuals that it is connected with, like an old-fashioned telephone switchboard with a tangle of wires connecting subscribers. By such an arrangement, the AM fungus–plant symbiosis has endured for hundreds of millions of years and it is still found in most plants. What is the secret of such spectacular success?

What has probably sustained this remarkable symbiosis for so long is the fact that despite their mutual dependence at a general level, both plant and fungus have alternative species of partner to choose from. Each has rivals that could gain advantage if they cheat, so cheating doesn't pay.[8] If this hypothesis is correct, then plants must have the ability to select among different fungal partners and to starve or reject them if nutrients are not forthcoming. On the fungal side of the bargain, they too must be able to reduce the flow of nutrients to poor cooperators and switch to other hosts. Experiments support these predictions.[9] Furthermore, plants can cut off their symbionts if they have an alternative supply of nutrients, for example from fertilizer, or when they are growing in shaded conditions that make a trade in scarce carbon a poor deal.

The AM fungal symbiosis may be resilient and ubiquitous, but it is not immutable. Plants in six different families exploit AM and other fungi as a carbon source, having lost their chlorophyll and the means to make their own food.[10] These ghostly pale plants have in essence become parasites on other, green plants, with their shared fungi in the common mycorrhizal network acting as a conduit. Furthermore, a sizeable minority, comprising about 30 per cent of all plant species, have swapped their AM fungi for other groups of fungi or, in the case of most of the cabbage family, made do with no mycorrhizal fungi at all.

Two big plant families, the orchids and the heathers, have swapped the original arbuscular mycorrhizas for their own particular brand of mycorrhizal system. Orchids are an especially interesting case because in this huge family of some 28,000 species,[11] all of them start life as parasites. The first five or six years of life for a terrestrial orchid, like a bee orchid or the common twayblade, are spent underground, parasitizing a fungus with nothing to offer in return until the seedlings finally get big enough to emerge from the soil and make a green shoot. Some orchids, such as the bird's nest orchid, a relative of twayblades, never break the parasitic habit of their early years and live their whole lives as leafless parasites. The Indian pipe, a plant in the heather family, is also a lifelong parasite of its AM fungus.

Orchids have dust-like seeds that are the tiniest and most ill-provisioned in the plant kingdom. Seedlings can only survive by recruiting fungal partners from the soil when the seeds germinate. This very chancy way to begin life probably only succeeds at all because suitable fungi are ubiquitous and orchid seeds are numerous.[12] A single orchid seed pod can contain millions of seeds. These are the tiny specks that you will find in ice cream flavoured with pods of vanilla, which is an orchid. Seedling orchids are undoubtedly parasites, but there is the intriguing possibility that the fungi that they parasitize are themselves obtaining food from the seedlings' parents nearby, in which case this could really be nursery care, not larceny. That would be some compensation for such shrivelled offspring. Against this idea is that orchid seeds are so light that most are probably carried by the wind far from their parents' reach.

The vast majority of orchids are green, so one might imagine that the frequency of cheating on fungi is low, but recent studies suggest otherwise. New techniques that determine how much of a plant's carbon, its staple food, comes from photosynthesis and how much comes via symbiotic fungi have revealed that there is a surprising number of species that dine somewhere between the green and the white.[13] Unlike AM fungi, at least some orchid fungi have two sources of food. There is the day job trading with green orchids and a night shift that requires no light because it involves feeding on dead wood. In one case, and there must be others, carbon acquired by a green orchid has been traced to its source in dead wood.[14]

The existence of plants that cheat on their mycorrhizal fungi suggests that, as in most symbioses, the relationship is not symmetrical. Plants are in control. This is demonstrated at the molecular level by a system of turnkey genes that plants use to admit fungal hyphae into their cells. Not all fungi are friendly symbionts—in fact most are parasites or live by decomposing dead material, so plants, like all hosts of symbionts, need to have a signalling system to distinguish friend from foe. Recent evolutionary analysis has discovered something fascinating about this system. All the relationships that plants have with symbionts that penetrate inside plant cells, as AM fungi do, use the same set of genes to admit friends and bar the gate against foes.[15] Orchids and heathers switched from AM fungi to other symbiotic fungi, but although the fungal partners are new, the genetic keys are the same as used for AM fungi. Land plants started out with the keys to the most intimate kind of intracellular symbiosis already made and these are still in use more than 400 million years later. Anyone who has left the code that unlocks their voicemail on its factory settings will understand how this happens. Inertia is a powerful thing in evolution too. What works is generally kept or modified, not abandoned or reinvented.

Admitting fungi into the inner sanctum of your cells is risky for a plant, even with a molecular system to distinguish friend from foe, because friends can become foes. This may be why many plants have evolved symbioses that replace the original AM system with one that does not require microbes to enter plant cells. In place of AM fungi that penetrate root cells, forest trees such as oaks, beeches, pines, and many others have a fungal symbiosis called an *ecto*mycorrhiza ('ecto' meaning outside) that performs the same function. The fly agaric, with its fairy-tale scarlet cap and white spots, and many of the other familiar toadstools of the autumn woods are the fruiting bodies of ectomycorrhizal fungi. Truffles are the subterranean fruiting bodies of fungi associated with oaks.

The roots of newly germinated tree seedlings are quickly sheathed in diaphanous fungal mycelium and will grow only poorly if this colonization by ectomycorrhizal symbionts is prevented. Ectomycorrhizal fungi penetrate between root cells, forming an internal net, but they do not enter inside cells or form intracellular arbuscules. Indeed, the trees that they associate with have lost the turnkey genes that would permit this.

Why have the genes that allow fungi to enter cells disappeared from ectomycorrhizal trees? When your houseguest has moved out, the hazard of still leaving keys under the doormat is obvious. So, while these genes had a beneficial function in symbiosis, they were preserved by natural selection, but when they were no longer needed, they were just a liability and were let go. This hypothesis is supported by the fact that one of the genes in question renders plants susceptible to invasion by a deadly pathogen called *Phytopthora*.[16]

In stark contrast to the single evolutionary origin of the AM fungal symbiosis, the switch to ectomycorrhizal symbiosis has occurred as many as 80 times in the last 200 million years.[17] So, while only about 2 per cent of all plant species are ectomycorrhizal, natural selection has repeatedly experimented with it. Ectomycorrhizal fungi are capable of extracting phosphorus from rocks. The need for this important nutrient may have driven plants to repeatedly team up with fungi that can supply it.

The same genes that plants use to control the entry of certain fungi into their root cells are also used to admit another important symbiont: the bacteria that feed plants in the pea family with nitrogen. The pea family, or the legumes, are the third biggest family of flowering plants after the orchids and the daisies and they seem to owe their success to symbiotic bacteria called rhizobia. Legumes are a relatively young family, starting out 60 million years ago, soon after the devastating meteor strike that killed off the non-avian dinosaurs.[18] Within a couple of million years, a mere blink of an eye in evolutionary time, the legumes had evolved their symbiosis with rhizobia. There is evidence that the ancestors of legumes had already evolved some precursor of symbiotic nitrogen fixation, giving them a head start, but exactly what that precursor might have been is another unknown.[19]

Rhizobia have the extraordinary ability to turn nitrogen from the atmosphere, where it is an inert gas comprising 78 per cent of the air, into a soluble form that plants can use for growth. Nitrogen is a major plant nutrient required by all plants, and indeed all life, but legumes have a private supply provided by symbiosis. It has been estimated that the nitrogen fixed by symbionts of legumes such as soybean grown in croplands and clover and alfalfa in pastures amounts to somewhere between 36 and 56 billion kilograms a year.[20] Wild legumes, of course, fix yet more. The industrial production of nitrogen fertilizer uses a huge amount of fossil fuel, accounting for 1 per cent of global energy consumption.[21] Symbiotic nitrogen fixation does the same job using only the power of plants.

As with the mycorrhizal symbioses, plants do not enter the world ready-equipped with their symbionts but must recruit them from the soil. Soils contain hundreds of species of rhizobia. Typically, these bacteria make contact with the very slimmest of roots called root hairs that in response curl around them, like a finger gesturing 'come-hither'. An exchange of molecular signals takes place and bacteria that pass the test are drawn into the interior of root cells that form a special chamber called a root nodule. Root nodules both accommodate and confine rhizobia, providing them with the specialized environment they need to fix nitrogen and preventing them from invading the rest of the plant. The parallel with the bacteriomes of insects is remarkable.

If you pull a bean or pea plant up by the roots at the end of the growing season, you will see hundreds of root nodules hanging from them like tiny baubles. And if you wash the soil off, you might notice that the nodules are pink in colour. The colour is caused by a plant protein called leghaemoglobin. This is the plant equivalent of haemoglobin that makes blood red, and it performs the same function, which is to transport oxygen. In blood, oxygen is transported around the body; in roots, it is transported into nodules for the use of the rhizobia inside. Why do symbiotic rhizobia need a special oxygen transport system? Other intracellular symbionts such as AM fungi don't have one. What's so special about rhizobia?

The answer is that rhizobia have to do two quite incompatible things with oxygen. On the one hand they are aerobic bacteria, and they need oxygen for respiration just like we do. On the other hand, they need to carry out the fixation of nitrogen using a process that will only work in the absence of oxygen. The solution to this conundrum is provided by the plant.[22] It surrounds the nodule in leghaemoglobin, which releases oxygen on demand only where it is needed. Incidentally, leghaemoglobin is chemically so similar to haemoglobin that 100 per cent plant-based burgers can be made to taste and appear meaty by the addition of leghaemoglobin.[23]

There is something else rather special and odd about rhizobia. There are many different kinds of rhizobia, but all have the same set of genes required for nitrogen fixation. It's as if you went to the zoo and discovered that it wasn't just the parrots and visitors who could talk, but that the lizards and lions also demanded, 'Who's a pretty boy, then?' How can unrelated species all share something as specialized and rare as the ability to fix nitrogen?

There are three possible answers to this question. First, the different rhizobia may all have inherited nitrogen fixation from a common ancestor. For example, all mammals are hairy because we share a common ancestor. This is conceivably the case for bacterial nitrogen fixation, but the different rhizobia are not all closely related, so the common ancestry solution must be rejected. The second possibility is that nitrogen fixation has evolved multiple times, as for example ectomycorrhizas have. However, biological nitrogen fixation only evolved once. All we are left with is the third possible answer, even though this sounds the least likely: the whole set of genes for nitrogen fixation must have spread as a unit between different, unrelated bacteria.

As Sherlock Holmes famously said, 'When you have eliminated all which is impossible, then whatever remains, however improbable, must be the truth.' The improbable truth is that nitrogen fixation genes move between different soil bacteria by a process of horizontal gene transfer, and they do this frequently.

Most genes, such as our own or those of plants such as legumes, are passed vertically down the generations from parent to offspring along a family tree. Horizontal gene transfer refers to genes that move across family trees, not from parent to offspring but from one unrelated individual to another. Such genes behave rather as viruses do, moving from an infected host to another, and some may indeed be moved by viruses. Horizontal gene transfer is very common in bacteria, and it is how the genes for multiple drug resistance spread among bacteria in hospitals.

Play in the legume team is getting complicated, so let's get everyone together down in the dugout for a review of the game. First, there is the plant, which definitely has a management role. The plant selects the rhizobia admitted to the team and hothouses them in nodules. Rhizobia only get to play if they are carrying the genetic credentials required to join the team, but the plant doesn't take credentials at face value. Performance in the nodule is the test and if a nodule doesn't deliver, it is let go. If this sounds familiar, it should, because it is an example of conditional cooperation.

Rhizobia are like Premier League footballers: they come from many different bacterial nations and what counts are the skills they bring, not where they hail from. Those skills come as a set and are transferable by horizontal gene transfer. Here the football analogy gets weird, and we must haul ourselves out of the dank and dirty dugout to see how the genes play on their own. A skilled footballer is in control of the ball, but rhizobia don't have charge over the genes for nitrogen fixation. The six or so genes that convert a soil bacterium into a rhizobium behave like a team within a team and they play only for themselves. These genes, which plug and play in different bacteria, are known as a cassette. It's the cassette and not its host rhizobium that decides when it's going to move on.[24] Think of a cassette as being like a pair of magic football boots that confers star-player status on whoever is wearing them. The genes in the cassette can been traced back to an origin three billion years ago in bacteria.[25] Didn't I say it was weird?

The favoured method a cassette uses to transfer to a different rhizobium is to induce the bacterium to have sex. Not so bad or so different from footballers after all, you might think. What's the advantage in the transfer market for the cassette? The likely explanation is that the cassette is in search of a better fit with the plant which controls how hospitably rhizobia are treated.

There is a chain of cooperation from cassette to rhizobium to plant and you might imagine that what is good for one link in this chain is good for all the rest. That would be true if all plants were equally good hosts for rhizobia and all rhizobia were equally good hosts for cassettes, but that is not the case. The reason is that each symbiont bears a cost as well as earning a benefit from cooperating, so of course there are cheats of various kinds seeking to freeload, necessitating checks and selection by plants.

Nitrogen fixation might seem like an unalloyed boon to any plant that has the genetic talent to team up with the right bacteria, but we know from looking at the evolutionary history of legumes and their relatives that this talent has been lost many times. In the plant, 30 different genes are required to make the symbiosis work, so it is a complex system. Complex systems are vulnerable in at least two ways. First, a process that depends on 30 genes is 30 times more susceptible to being disrupted by a random mutation knocking out a gene than a process depending only on one. And second, genes are often involved in multiple jobs, so a complex process involving 30 genes is quite susceptible to evolutionary pressures working on other tasks that some of those genes do. This could mean that preserving nitrogen fixation exerts a cost to the plant in some other function that is important to its survival and reproduction. If a symbiosis becomes costly, cooperation will cease as mutations that destroy the genes it once depended upon are favoured or not selected against. Disabled symbiosis genes stand silent witness to this happening in the genomes of many plants whose ancestors once fixed nitrogen.[26]

What is the take-home message from root symbioses, or what might fancifully be described as these phytosympathies? Simply, it is that cooperation can be many layered, dependent, and complicated, as it is in the rhizobium–legume symbiosis, and stable, as it is with AM fungi, without the occurrence of any MTE. All the partners in these symbioses have a choice of partners and the opportunity to select a new one each time an uncolonized seedling appears. This situation sustains cooperation without the teams involved becoming bound for ever into a new entity. As a result, plants can change symbiotic partners quite radically over time, as their evolutionary history reveals they often have.

We are not quite done with phytosympathy. The common mycorrhizal network allows flows of nutrients and signals between the plants connected into it. Do the trees connected in this way cooperate with each other? Can trees be friends?

Let's take a walk in the woods. We are in the subalpine zone, about 1,200 metres up Whiteface Mountain in the Adirondacks of upstate New York. New York City lies nearly 300 miles due south of here. Once, before the opening up of the West, like most of New England this was farming country but the rugged, infertile land here in the East could not compete with the productivity of California and so huge tracts of land were abandoned and spontaneously rewilded by nature. Leisure has long been the main economic activity in the Adirondacks and Whiteface Mountain is a tourist attraction with a road winding to the top that was built in the 1930s, during the New Deal era. But we are not here for the view or the skiing, which is just as well because at this elevation the forest is so dense that both are impossible. Entering the forest, the dark foliage of balsam firs creates a deep gloom, dripping with moisture from the clouds that so often cling like a mantle over the slopes.[27]

In this demanding subalpine environment, with fierce icy winds in winter and shallow soil, there are no very old or very tall trees. Nothing is more than two storeys high, to use an urban unit of measurement that seems apt for New York State. As we press through the firs, their springy branches armed with sharp twigs slap back at us, as if trying to bite. Then we come up against a thicket of spindly saplings, all stem and no leaves and not much taller than we are. Their stems are so close together that we can only progress through them with gloved hands outstretched to part the prickly stalks, as though diving into a perpetually oncoming wave. Then, suddenly, the thicket gives way and we find ourselves in a clearing. This is what I have brought you to see. The clearing reveals a band of open sky, stretching away to either side of us. It is occupied by tall, dead trees, many still standing, some leaning upslope as the wind has felled them, and others with trunks snapped off halfway up. Beneath this strip of dead trees is a carpet of fir seedlings, bathed in light.

Almost nothing grew on the floor of the forest until we reached this glade, but here the ground is alive with new life: up to 3,000 seedlings per square metre. These are the offspring of the mature and now dead firs in this gap, whose parting gift to their progeny was a clear view of the heavens. From the two-storey trees at the start, through the whippersnappers and the thicket of striplings to the seedlings now at our feet, we have just traversed a fir wave. This is a very rare forest formation, in which a peculiar combination of topography, climate, and tree biology causes the tallest trees of the forest to die in strips lying at right angles to the direction of icy winter winds. Each winter, the wave of death advances another step into the forest, killing the tallest, most exposed trees. In spring, the light that follows the winter slaughter ushers in a new generation of small trees. Over time, these saplings grow and compete with one another, and the survivors go through each of the stages we have ourselves walked along. By 80 years of age, only one tree has survived from a cohort of 10,000 seedlings.

The life and death that we have witnessed in a fir wave is what happens in all forests, but only in a fir wave are all the different stages of the life cycle lined up in chronological order for us to see so clearly. In other woods, the glades where the forest regenerates are small, scattered patches of light, not wave-cut strips in the canopy. All trees, indeed most organisms, produce a superabundance of offspring, only the tiniest fraction of which make it through to adulthood. Malthus would have recognized the process at Whiteface Mountain and Darwin would have hailed it as a perfect example of the struggle for existence. Kropotkin, I suspect, would have looked away. The tallest seedlings capture the light, the rest are shaded and die of light starvation. It's grow or die. So, can trees be friends? On this evidence you might think not. And yet.

And yet, the struggle for light that turns foliage into carnage is not the whole story, because below ground is the common mycorrhizal network. Mr Kropotkin, you can unshield your eyes. Meanwhile, in another part of the forest, in fact on the other side of North America back in the 1990s, a PhD student called Suzanne Simard is standing in a very different forest among stately Douglas firs, Western red cedars, and birches. She is wondering whether seedlings that are connected by the common mycorrhizal network might be sharing carbon and thereby helping each other. To test this, she plants out young trees of the three species and, once they have established, she feeds their leaves with carbon dioxide labelled in such a way that she will be able to trace carbon that has been absorbed by one seedling and transmitted to another. The results show that carbon has in fact moved between fir and birch, which share ectomycorrhizal fungi, while almost none of the labelled carbon shows up in cedar seedlings. Cedars have AM fungi and are not connected to the other two species. The results support the idea that birch and fir help each other via their shared mycorrhiza, and Simard publishes the results in *Nature*, the world's foremost scientific journal.

Suzanne Simard's study came out in 1997 and was heralded by *Nature* as demonstrating a wood-wide web connecting trees below ground. As sometimes happens with studies that vault with such fanfare from the pages of *Nature*, Simard's work attracted scepticism from some quarters, which she found painful.[28] Twenty-five years later, Simard's research and her personal story have inspired novelists, film makers, and others. The scientific response has been more measured. Some other studies have given qualified support to the sharing of carbon and nitrogen over common mycorrhizal networks in mature trees as well as seedlings,[29] although this is by no means universal.[30] From the start, Simard herself was at pains to say that sharing only happens under certain conditions and in this she was correct. One of the necessary prerequisites to resource sharing is that of course seedlings have first to survive the massive mortality that happens in early life.

As to why trees share resources, we don't really know, but the general conditions for cooperation have already been well rehearsed and may contain the answer: kin selection and interdependence. However, why the *trees* move carbon may actually be the wrong question. One of the early criticisms of Simard's study was that the movement of carbon is under the control of the fungi, not the plants, and that fungi may move it around to get the best return for themselves from their symbionts.[31] So maybe plants are not actively helping each other at all, they are just the passive donors and recipients of fungal largesse. This could explain why resources are generally moved across the mycorrhizal network from trees supplying a lot of carbon to those with less where the AM fungus would need it more.

The discussion of the common mycorrhizal network and its popular manifestation, the wood-wide web, is not the whole story of how plants cooperate, and indeed it may not even be the most important way that they help each other. Trees of the same species, but sometimes of different species, are often directly connected to each other by root grafting. This stabilizes all the trees against windthrow. Transport of carbon from dominant to shaded trees through root grafts has been observed[32] and this may be of mutual benefit, or it might simply be the unavoidable cost of obtaining physical support. Grafts to living trees can keep the root systems of felled stumps alive for a decade or more, which has been interpreted as an altruistic act.[33] However, the benefit to the surviving tree seems obvious—it has inherited an extension to its root system.

In dry habitats, it is common for deep-rooted trees to draw water to the surface during the night and to leak this out during the day, benefitting neighbouring plants of other species.[34] At first sight it is difficult to see how this benefits the tree supplying the water, but it may do so indirectly if the vegetation around the tree and the soil beneath it become reservoirs holding soil water.[35]

In general, as Kropotkin observed in Siberia, adverse conditions favour cooperation. Experiments comparing how plants growing in conditions of high or low stress respond to the removal of neighbours have found a consistent pattern: at high elevation in mountains such as where fir waves occur, plants do worse when their neighbours are removed. At low elevations, where the climatic conditions are more favourable to growth, plants grow better when their neighbours are removed.[36] The conclusion is that in stressful environments, plants help each other to survive and grow. Plants tend to be fair-weather enemies but can be ill-weather friends.

So, what is the conclusion to the question, 'Can trees be friends?' My answer is: 'Neighbours certainly, but friends only in the right circumstances.' This also neatly sums up how humans acquired their best friend.

8

Good companions

It is an arresting sight. A man sitting on a park bench with a grey wolf in his lap. Both are clearly contented in each other's company. The wolf is a mere 25 centimetres long, though fully grown. It is a domestic dog, of course, but sometimes science makes you see things through a different lens. Despite their huge variety, from diminutive Chihuahua to giant mastiff, all breeds of dog descend from the wild grey wolf, which first teamed up with humans 30,000 years ago. There were probably two domestication events, but the wild progenitor was the same, pack-hunting carnivore in both cases.[1] Dogs were the first animal to be domesticated by our ancestors, and since then we have made a dairy cow out of the aurochs that was so ferocious it scared the Romans, turned wild pigs into the magnificently corpulent Empress of Blandings,[2] transformed wild stallions into Black Beauty, bred brightly decorated koi from muddy wild carp, and fashioned chicken nuggets from jungle fowl.

The process by which we have changed once-wild species into domesticated animals and plants is merely the artificial counterpart of natural selection in the wild. Through artificial selection and selective breeding, we have made animals tame, plants pliant, and both productive. 'Tame' means cooperative with humans and disinclined to cheat on the bargain by natural aggression or attempting to escape captivity. However, domestication is something more than mere tameness, which can be induced in wild animals through training. A domesticated animal or plant is genetically altered. By passing the genes of wild species through the eye of a domestication needle, we have sewn the fate of these animals and plants into our own. Nearly all of this happened long before the science of genetics was known. The earliest of these relationships, between human and dog, now form cooperative teams to hunt, to herd livestock, and to run truffles and fugitives to earth.

Using their extraordinary sense of smell, dogs can even be trained to detect illness in humans.[3]

We can't be sure how the first wolves became associated with humans, but they are carnivores, and it would be natural for them to scavenge around hunting camps. The animals least afraid of people would have approached nearest and got the best pickings, starting the selection for tameness and camp-following. Wolves are cooperative breeders and highly social, which may have predisposed them to team up with humans. Wild wolves are feared, but at some point people began to find their presence useful, perhaps because they acted as sentinels around the camp,[4] which is another example of the counter-intuitive tendency for antagonism to give birth to cooperation.

Though wild wolves were to be found throughout the northern hemisphere, it was the descendants of just the two domesticated populations that became our travelling companions wherever our hunter-gatherer ancestors went, reaching from the Arctic to Australia. The long association between humans and dogs can be divided into three phases during which we shaped dogs in different ways. A tantalizing glimpse of the relationship in its first phase is to be found in the deep recesses of a cave in the Ardèche gorge in southern France. The walls of the cave, discovered by Jean Marie Chauvet and two companions, are decorated with Palaeolithic paintings of aurochs, bears, lions, and other large animals that were hunted at the time, some 20,000 years ago.[5] People blew red ochre upon the walls, airbrushing the stencilled outlines of their outstretched hands on to the surface, as if staking a claim in posterity with an unmistakable signature of their humanity. No other signs of humans and none of dogs is present, except for one stunning tell-tale.

In the soft limestone sediment at the very back of the cave are the footprints of a child, walking with unhurried steps. Alongside the child's footprints, as if walking by her side, are the paw prints of a fully grown wolf.[6] Who can say what these tracks, hidden for 20,000 years, mean? Was this a child and a domesticated wolf behaving like members of one family, or is the coincidence of footprints merely that: a coincidence between tracks made at different times? We cannot know, but coincidence or not, these prints in the Chauvet cave symbolize the first phase of a long relationship. We can confidently say two things about this phase of 20,000 years following initial domestication: hunting bound hound and human together and during the period, dogs diversified into at least five major lineages, reflecting human migrations.[7] Then everything changed with the invention of agriculture.

Agriculture appeared near-simultaneously around 10,000 years ago in over 20 locations around the planet, marking the start of the second phase of the dog–human relationship.[8] The edible bounty produced by farming propelled humans and our canine companions to an abundance that changed the face of the Earth and the faces of dogs. By one estimate there are a billion dogs on Earth today.[9] With agriculture came settled communities, surpluses of food, cities, new occupations, and hierarchies of power and wealth. Dogs were now useful in a greater variety of roles and were selected accordingly.

In Mesopotamia, where the first cities were created, hunting dogs appeared on gold jewellery 5,000 years ago. A thousand years later, a figurine of a large dog resembling a modern mastiff was fashioned bearing the inscription, 'Don't stop to think. Just bite.'[10] 'Man bites dog' is bruited as the epitome of newsworthiness, but it was very old news to the Olmecs of southern Mexico. Three thousand years ago, Olmec farmers paid their taxes in dogs fattened for the table on an exclusive regimen of corn, which would have been an unfeasible diet for their wild ancestors.[11] The Aztecs bred the ancestor of the Chihuahua as a lap dog.

Writing in the first century, the Roman author Columella explained why different kinds of dog were desirable on a farm:

> Buying a dog should be among the first things a farmer does, because it is the guardian of the farm, its produce, the household and the cattle. It should be big and have a loud bark, first to intimidate the intruder when it is heard and then when it is seen. Colour too is important. An all-white dog is recommended for the shepherd to avoid mistaking it for a wolf in the half-light of dawn or dusk, and an all-black guard dog for the farm to terrify thieves in the daytime and be less visible to trespassers at night. The farm-yard dog should be heavily built, with a large head, drooping ears, bright eyes, a broad and shaggy chest, wide shoulders, thick legs, and short tail. Because it is expected to stay close to the house and granary, its lack of speed is not important. The sheepdog, on the other hand, should be long and slim, strong and fast enough to repel a wolf or pursue one that has taken its prey.[12]

Dogs were not just farm animals though. After the transition to agriculture, human settlements everywhere acquired a population of feral dogs that to this day live around human habitations in the Global South. These 'village dogs' still live in groups, but passage through domestication has changed their social behaviour. Whereas the members of a wolf pack are kin and only the dominant pair breed, among village dogs there is no hierarchy and all animals breed. A study of feral dogs in India found that females gave most care to their own pups but would care for others too.[13] The genetic related-ness within the groups was unknown, but the breeding system is probably best described as communal.

The third phase of dog domestication had as big an effect on dog evolu-tion as anything in the previous 30,000 years and can be dated very precisely to a particular historical event: the founding of the Kennel Club in London on 4 April 1873. The purpose of the founders of the Kennel Club was to establish rules that would standardize the judging of different breeds at dog shows and to preserve the breeds that they considered ancient. An American Kennel Club was formed in 1887, operating along similar lines. The Kennel Clubs designated a number of different breed classes, such as Gun Dogs (retrievers etc.), Hounds (bloodhound, greyhound etc.), and Toys (Chihuahua etc.).

Each breed in each class has a very detailed standard which specifies what the ideal dog of the breed should look like and how it should move and behave, as defined by 17 characteristics. Each characteristic has further mul-tiple criteria, so for example the forequarters of a golden retriever should have '[f]orelegs straight with good bone, shoulders well laid back, long in blade with upper arm of equal length placing legs well under body. Elbows close fitting.'[14] These stringent criteria have channelled dog breeding to such a degree that an evolutionary tree of modern dogs reproduces the classes and breeds of the Kennel Club standards as though they were what nature created over millennia, even though they are the product of artificial selection mainly during the last 150 years.[15]

As the most senior member of the domesticated fauna, the dog is a legacy of our hunter-gatherer past. By contrast the domestic cat, our other com-panion carnivore, is a totem of what came next: agriculture. Evolutionary analysis reveals that cat domestication began in the region known as the Fertile Crescent of South-west Asia, most probably around 11,000 years ago.[16] This is the place and time when cereal agriculture took off. A com-bination of archaeological and genetic evidence for people, wheat, house mice, and cats paints a picture of what happened.

There is a domestication syndrome in plants that, like the syndrome in animals, occurs in all grain crops when they come under artificial selection. The syndrome is mainly visible in seeds which, as domestication proceeds, become larger and lose the dormancy that most wild species growing in seasonal environments have. Also, wild grasses disperse their seeds when they ripen, but domesticated cereal grasses such as wheat and barley do not. A single gene determines whether wheat seeds split off naturally from the seed head, a process called shattering, or stay attached until physically broken off when the grain is threshed. This difference is visible in seeds excavated at archaeological sites across the Fertile Crescent and it shows that wheat domestication was a gradual process that began thousands of years before the dawn of farming 11,000 years ago.

Between about 20,000 and 11,000 years ago, people in the Fertile Crescent were gradually shifting their subsistence from hunting and gathering to living in more permanent settlements and farming. House mice seem specifically adapted to permanent dwellings and first appeared in human habitations in the area 14,500 years ago.[17] There are many wild species of rodents that live on the seeds of wild plants, and these will enter dwellings on occasion, as I discovered once while on a field trip to the Chihuahua Desert in Mexico when I was woken in the night by a kangaroo rat scampering up my leg in the darkness. Normally though, even in Mexico, far away from the Fertile Crescent, a mouse in a house is a house mouse.

By 11,000 years ago settled agriculture was established in the Fertile Crescent and house mice lived in houses and granaries, feeding on grains. Wild cats would have been attracted to human habitation by the abundance of mice, starting the process of cat domestication. By 10,500 years ago the farmers who colonized the Mediterranean island of Cyprus brought with them from the nearby Fertile Crescent not only crops but also house mice and cats.[18] Presumably the mice were stowaways, but the cats were deliberately introduced. In historical times, cats were carried on ships to control rodents and by this means were dispersed around the world. Wherever they went, cats displaced more local animals that were kept for rodent control. In Ancient Rome and Greece, ferrets (domesticated polecats) were kept in houses to control mice and in China a local wild feline provided the same service, but both were eventually replaced by the advance of the domestic cat from the Fertile Crescent.[19]

Domestic cats provide an interesting contrast to domestic dogs, reflecting the different social behaviour of their respective wild progenitors. Grey wolves are cooperative breeders and hunt in packs, while wild cats are territorial, solitary hunters.[20] Domesticated cats living as singletons in the home are perforce not especially social, but the social behaviour of feral domestic cats depends on whether or not they have to hunt for food as wild cats do. Feral cats in wild places with a low density of prey, as for example in rural Scotland, are solitary like their wild ancestors. However, where there is food in abundance, such as on farms or in cities, feral cats form matriarchal groups of related females that will indiscriminately nurse each other's kittens and defend their territory from other groups. These groups of cooperative breeders are more like female lions in their social behaviour than wild cats, though unlike prides of lions, groups of feral cats have no dominant male.

Cats are the only animal to have changed from a solitary existence in the wild to sociability under domestication.[21] One view is that solitariness made the wild cat an unlikely subject for domestication, but this assumes that the wild species is an inflexible loner, and maybe this is wrong. Feral cats are flexible in their social behaviour and notoriously difficult to distinguish from wild cats. The puzzle of how an unsociable animal was domesticated evaporates if we allow for the possibility that cat social behaviour is inherently more flexible than we have given it credit for.

The typical domestic cat is more independent of its owner than the typical dog: it hunts for itself, visits other humans for food, and chooses its own mates on nocturnal prowls if not neutered. Perhaps the wild cat is only semi-domesticated? A comparison of the genomes of wild and domestic cats supports this idea, showing some changes believed to be signatures of domestication but fewer than are found in dogs, which have been domesticated for much longer.[22]

While dogs and cats were first domesticated for the practical assistance that they could render us, these relationships are no longer just transactional, and most of these animals are now kept as companions. The connection between people and their pets is emotional. Dogs, cats, and some other domesticated animals are able to read human intentions and emotional states and recognize their owners.[23] It has just recently been discovered that dogs have novel facial muscles not found in wolves that enable them to intensely raise their eyebrows, creating the appearance of puppy dog eyes when looking at humans.[24] Symbiotic microbes may get inside cells, but domestic pets get inside our heads.

Domestication is as much an ecological process as an evolutionary one, both in its causes and its consequences. The animals and plants that we domesticated are the ones that we depended on for food in our local environment, or for hunting and pest control. The ecological consequences of domestication are so large that they cannot be overestimated. Without agriculture there would be no cities, no industrialization, and no anthropogenic climate change. Agriculture was the midwife of the Anthropocene. In the groundbreaking book *Guns, Germs and Steel*, Jared Diamond shows how momentous historical events can be traced back to the consequences of domestication.[25] The germs of the title cause the zoonotic diseases that spread to humans from animals, most often domesticated ones. By some estimates, one-third of the human population is infected with a particular disease of cats.

The disease toxoplasmosis is caused by an intracellular parasite called *Toxoplasma gondii*, which reproduces sexually only in cats and is passed on in cat faeces. Other mammals, including humans, can become infected from contact with cat faeces or contaminated vegetation and soil. In these mammals, the parasite forms cysts in the central nervous system, including the brain. In lab studies, infected rodents such as rats and mice change their behaviour, showing less fear of cats, which probably increases the likelihood that the rodents will be eaten.[26] A long-term field study in Kenya found that hyena cubs infected with *Toxoplasma gondii* approached closer to lions than uninfected cubs and more often fell prey to the big cats as a result.[27] Like many parasites, *Toxoplasma gondii* appears to be able to alter the behaviour of infected animals in a manner that improves the likelihood it will reach a feline where it can reproduce. Humans are rarely eaten by felines, so they are a dead-end host for the parasite, but *Toxoplasma gondii* can still alter human behaviour. There is evidence that people infected with the parasite have an elevated risk of several mental disorders, particularly schizophrenia.[28]

Although domestication is normally taken to be a human action, as we have seen it is not always deliberate and therefore it would be arbitrary to ignore analogous processes in other species. The distinction between symbiosis and domestication can be a fine one, for example in lichens. Most of the photobionts in lichen symbioses are also found free-living and so cannot be considered domesticated, but *Rhizonema* are cyanobacteria that are found only in lichens.[29] *Rhizonema* occur in symbiosis with different, unrelated lichenized fungi, indicating that they have moved from one to another by horizontal transmission, like the seeds of a plant variety moved between different farmers and improved along the way. Lichens have been described as the 'fungi that discovered agriculture'.

If lichens are the fungi that discovered agriculture, then leafcutter ants are the insects that discovered fungiculture.[30] Leafcutter ants found in the forests of tropical America harvest fragments of living leaf from high up in the tree canopy and transport these down the trunk and across the forest floor to a huge nest. The leaves are not consumed directly by the ants but are fed to a fungus that they cultivate within the nest, and which produces protein-rich growths that the ants eat (Plate 8). When a new nest is formed, the founding queen ant inoculates it with a piece of fungus from an existing nest in the same way that a farmer might sow a potato field with tubers from a previous crop. In fact, the fungus has been domesticated by ants in the same way that we have domesticated our crops, but ant fungiculture has been going for more than 50 million years, and some fungus clones, carried from nest to nest generation after generation, are 20 million years old.[31] An experiment that swapped the fungi between two closely related but different species of leafcutter ant revealed that the ants and the fungi both did worse than when growing with their customary partners.[32] Long association has allowed evolution to hone the fit between team members to their greater mutual advantage.

As the first animals to be domesticated, it was in dogs that a set of traits known as the domestication syndrome first appeared. Under domestication, dog's bodies and brains became smaller, their ears floppy, their tails curled, and their coats often piebald. None of these traits was directly selected for, but dogs and many other animals selected by humans for tameness express this domestication syndrome.

Why should selection on behaviour produce such a strange set of anatomical changes that recur in pigs, dogs, cattle (Figure 13), sheep, horses, and other domesticates? Charles Darwin spotted the pattern and puzzled about this question in his two-volume book *The Variation of Animals and Plants under Domestication*, but he was groping in the dark in an age before the development of genetics.[33] Indeed, it was in the hope of understanding inheritance that he laboured, writing a thousand pages containing every fact on biological variation in domesticated animals that he could lay his hands on, finally emerging with a hopeless theory called pangenesis.

It is an irony of scientific history that Darwin, like everyone else at the time, failed to appreciate the significance of Gregor Mendel's breeding experiments with peas that had been published in 1866 and which later became a foundation stone of genetics. Reading Darwin describe in his *Domestication* book the results of his own crossing experiments with flowers, one can see with the benefit of hindsight how tantalizingly close he came to results like those obtained by Mendel, but alas how far he was from

Figure 13 A domesticated zebu bull, *Bos indicus*

understanding their meaning as Mendel had done. History was to prove that to explain the underlying cause of the domestication syndrome that Darwin discovered would not just require genetics but big advances in embryology as well.

All the right ingredients for a viable explanation of the domestication syndrome did not come together until 2014, when Adam Wilkins, Richard Wrangham, and Tecumseh Fitch came up with an hypothesis based in developmental biology. They suggested that the many apparently unconnected traits of the domestication syndrome were all in some way related to tissues whose development is influenced by a region of the vertebrate embryo called the neural crest.[34] As already mentioned, the sundry traits of the domestication syndrome are not themselves the target of selection by animal breeders. No one said to themselves, 'Wouldn't it be fun if my pigs had floppy ears and a curled tail just like my dog', and then wasted time on realizing this whimsical notion. What certainly did happen is that there was selection for tameness.

The evolution of tameness depends on a reduction of fear and stress, which are both controlled by hormones produced by the adrenal glands—a pair of organs that sit on top of the kidneys. Both domesticated and experimental animals selected for tameness have adrenals of reduced size. The size of the adrenals themselves has no effect on the various traits of the domestication syndrome, but guess what? Development of the adrenal glands is affected by the number of cells in the neural crest. The hypothesis advanced by Wilkins et al. is that selection for tameness favours animals that have small adrenals and that this is a natural consequence of those animals having a reduced number of cells in the neural crest during embryo development. The reduction in the number of neural crest cells affects all the developmental pathways in which those cells are involved, notably brain growth, facial and tooth development, the melanocytes that pigment the skin, and cartilage in the tail. This would be how selection for tameness produces such an odd collection of side effects.

This ingenious hypothesis would doubtless have delighted Darwin, who would have been just as astonished to learn that it has been tested using genetics. A study homed in on the sequences of 11 key neural crest genes to look for evidence that they had undergone recent selection in domesticated animals.[35] Fifteen different domesticated species including house mice, rabbits, dogs, and camels were compared with their wild relatives, revealing that for most of the genes there were changes consistent with artificial selection. This is not surprising, you might think, given that we know that the animals in question were domesticated. However, 11 other genes chosen at random and with no connection with the neural crest did not show the same signature of selection in domesticated animals. This study therefore supports the neural crest hypothesis, but there is more.

We have surrounded ourselves with domesticated animals and plants that have been genetically chiselled to human ends, but is it possible that we domesticated ourselves as well? In 1872, the English social theorist

Walter Bagehot wrote that 'Man...was obliged to be his own domesticator; he had to tame himself.'[36] Bagehot thought that self-domestication would arise from group selection, but we have already seen that natural selection can explain cooperativeness in humans and that group selection is a less likely explanation because it is easily undermined by cheats.

In 1871, in his book *The Descent of Man*, Charles Darwin suggested that sociability would evolve from a starting point in what we would now call cooperative breeding:

> The feeling of pleasure from society is probably an extension of the parental or filial affections, since the social instinct seems to be developed by the young remaining for a long time with their parents...the individuals which took the greatest pleasure in society would best escape various dangers, whilst those that cared least for their comrades, and lived solitary, would perish in greater numbers.[37]

This intuition of Darwin's is one of many examples of astonishing prescience in his work. It was not until 2014, 144 years later, that a comparative study showed that 'filial affections', or the extent of cooperative breeding, correlates very closely with the degree of prosociality in different primates.[38] We know that social behaviour evolves and that in our own and other species this has produced cooperativeness among both kin and non-kin, but in our own case should we regard this as self-domestication? It's certainly a thought-provoking claim, but does it mean more than simply selection for cooperation? In the view of its proponents, it does.

Anthropologist Joseph Henrich sees the impulse for self-domestication arising from culture:

> Thus, cultural evolution initiated a process of self-domestication, driving genetic evolution to make us more prosocial, docile, rule followers who expect a world governed by social norms monitored and enforced by communities.[39]

By this account, human self-domestication is just as much a cultural product as is the domestication of plants and animals. Do humans show the biological and genetic signatures of the domestication syndrome seen in other animals?

Of course, there is no curled tail, nor floppy ears or piebald colouring, but in our recent evolutionary history there is a slight reduction in cranial capacity and a feminized face.[40] What of the genetic evidence? There are roughly 25,000 genes in the human genome, a small minority of which show a tell-tale signature of recent change due to natural selection. A study compared these changed genes with those known to have been under selection in domesticated dogs, cats, horses, and cattle to identify whether they were the same ones.[41] If humans are self-domesticated, then you might expect a significant similarity between the genes that have been under selection in humans and the ones under selection in domesticated animals.

Forty-one genes were found to have been under selection in humans and in one or more of the domesticated species—a highly significant overlap. This suggests a general similarity in the evolutionary forces acting on humans and the animals we have domesticated. However, only 1 of the 41 genes that changed in humans also changed in *all four* of the domesticated species, which suggests that the similarity is limited, even between different domesticated animals. One should also ask whether the selective forces in question were necessarily due to domestication. They might not have been, since all species have been exposed to other changes in the last 10,000 years, notably climate.

What does this all add up to? Tameness is the one trait shared by all domestic animals and humans. Of the various other traits that have been included in the domestication syndrome, such as decreased brain size or floppy ears, these all vary between domesticates.[42] The neural crest hypothesis seems promising as an explanation of the link between selection for tameness and other traits, but as ever in biology we would expect variation among species, which is what we see. There are parallels between animal domestication and human selection for prosociality, but there are differences too.

Does domestication lead to MTEs? Major transitions take place in two stages. The first involves team formation and the second is transformation when the reproduction of the team members becomes subsumed into the reproduction of the team. At that point, a new kind of individual is created. The plants and animals domesticated by humans can all be said to be members of a team whose fate is brought into alignment with human wants through artificial selection. But there are no instances where the products of human domestication, however valuable they may be to us, are sewn so tightly into our reproduction that an MTE has taken place. Dogs are not mitochondria, or even close.

If one wants to consider the lichen symbiosis with *Rhizonema* or the relationship between leafcutter ants and their fungus as cases of domestication, then these have transformed into new kinds of individual and they therefore do constitute MTEs. However, this transformation is really a product of symbiosis rather than domestication per se. Arguably, nothing is added to our understanding of lichens, leafcutter ants, or evolution by the analogy with human-domesticated crops. It is instructive to ask the same question about human self-domestication: has it led to an MTE?

For the answer to be in the affirmative, cooperation between individuals would have to have evolved further than cooperative breeding, towards something like eusociality, as it has in naked mole rats for example. That has not happened and does not seem likely to do so. But a different answer is possible if one uses a different definition of what constitutes an MTE. The definition that I have been using is that now favoured in evolutionary biology, which requires a new kind of individual to emerge.[43] However, this is narrower than the one originally proposed by Maynard Smith and Szathmáry in 1995.

They emphasized that when there is a major transition, the mode of transmission of information changes. Because genes carry information, the examples we have discussed so far do fit this definition. Genes and the information they carry are packaged in a new way when lichenized fungi and their photobionts are bundled together, or when the bacteria in an insect's bacteriome are transmitted through the insect's eggs, forming new kinds of individual. The broader, informational definition and the narrower individuality one that I have used produce essentially the same picture until we get to discussing culture and language.

Culture and language evolve, but these are built upon biological foundations (the brain, social behaviour) without themselves being biological. So, did human evolution and self-domestication produce an MTE when language appeared? Using the information definition of an MTE the answer would be yes because language introduced a new means of transmitting information, but using the individuality definition the answer is no.[44] By the latter definition, language is an emergent phenomenon that arose from how humans in groups communicate. Those human groups are not individuals in their own right. Groups don't reproduce, people do.

In our journey through the heights of cooperation to the depths of the history of life we have now reached another major transition. The individuals we have been discussing are made up of cells. How did cells get to form such teams, and how do they cooperate with one another?

PART III

Cells

9

A brand-new bag

James Brown, the godfather of soul music, got it exactly right when he belted out in his inimitably raucous voice, 'He ain't no drag, Papa's got a brand-new bag'. It *was* no drag when life got a brand-new bag. It was the Grandpapa of all MTEs. A cell is a magician's bag that when brand new, around four billion years ago, captured the essential molecules of life and contained them so that they could perform chemical reactions together. The first cell must have been a flimsy, membranous receptacle containing only simple molecules in a saline solution. But we don't really know, because even the simplest of living cells today is a bag of very sophisticated tricks, far removed from its ancestral cell. The earliest cells doubtless disappeared when their more sophisticated descendants ate them, or just stole their lunch.

The cell's greatest feat is to turn nutrients and energy into a new copy of itself, which the fastest bacteria can do about every 20 minutes. Imagine a party magician who pulls out of his bag not an endless stream of silk scarves or a live rabbit but a perfect copy of himself carrying a bag from which this doppelganger draws a copy of himself carrying a bag from which he draws...until all resources have been consumed and it is wall-to-wall members of the Magic Circle. That is some trick, and without its cellular equivalent there would be no life. How life began is uncertain and is explored later, but we do know what must have happened after the first bacterial cells appeared: they began to diversify and team up.

To be small is to be at the mercy of the environment, but to be a bacterium is to never be alone. Alone, a miniscule bacterium consisting of one lonely cell could scarcely survive, but in a colony of millions, it is the environment that is at the mercy of bacteria because bacteria team up. The earliest evidence of this is in rocks called stromatolites, found where the geology of the Precambrian of 3.5 billion years ago is exposed. In cross-section, these rocks reveal that they were laid down in layers, like domes of mille-feuille

Figure 14 Stromatolites

pastry piled high in a Parisian patisserie. The layers are thin mats made of bacteria and trapped silt soldered by slime. We know how they were made because stromatolites still form in the warm shallow coastal waters of the Bahamas and at Shark Bay in Western Australia (Figure 14).

Modern stromatolites, probably like their Precambrian ancestors, grow as year by year a new layer of photosynthetic cyanobacteria coats the surface like a fresh lick of paint. Many other species of bacteria colonize the mat: some are tenants and pay rent in the form of nutrients; others are squatters and only take shelter. Everywhere that microbiologists wield a swab, the power of DNA sequencing reveals bacteria in crowds, in cahoots, and at war. Bacteria are social beings—something that Peter Kropotkin uncannily anticipated well over a century ago in his book *Mutual Aid*:

> [W]e must be prepared to learn some day, from the students of microscopical pond-life, facts of unconscious mutual support, even from the life of micro-organisms.

Among the social accomplishments of bacteria are the exchange of nutrients and an ability to sense the abundance of their own kind and to respond accordingly. Some social interactions are cooperative, others are cheating or competitive, and some are lethal.

Lacking the ability of more complex cells to engulf and devour, bacteria interact with each other via the molecules that they secrete into the environment. This may sound like a handicap to cooperation, but the repertoire of social interactions is a rich one, even though the fundamentals are simple. A secreted molecule may be either beneficial or antagonistic to other bacteria. Which it is will depend on the identity of the producer and the receiver, as well as the nature of the molecule and the environment. Both beneficial and antagonistic interactions may be put to practical use.

Bacteria secrete into their environment enzymes that digest proteins and carbohydrates and convert them to smaller molecules such as amino acids and sugars that they can absorb. This can have a practical use in cleaning. In a somewhat gruesome case, a gel containing bacteria was recently used by art restorers in Florence to clean the marble tomb of Allessandro Medici, sculpted by Michelangelo in 1520. Allessandro was assassinated and interred in the marble sarcophagus without the usual mortuary practice of eviscerating the corpse. The ensuing products of decomposition leaked through the marble and stained it badly. Restorers tried various bacteria on the stubborn stains, but the most effective was *Serratia ficaria*, better known as a cause of urinary tract infections. This bacterium removed the 500-year-old stains in two days.[1] How many dry cleaners can do that?

Antagonistic interactions can control harmful bacteria in fermented foods. Lactic acid bacteria, like the ones used to ferment milk into yoghurt and cheese, produce small molecules called bacteriocins that kill other bacteria, including the food-borne pathogen *Listeria monocytogenes*. Bacteriocins are chemical weapons used in microbial competition for resources and may be narrow or broad in their range of targets. *Lactobacillus acidophilus* and *Streptococcus thermophilus*, which you may find name-checked on the labels of yoghurt pots, both increase the production of bacteriocins in the presence of *Lactobacillus bulgaricus*, which competes with them.[2]

When useful but costly molecules such as enzymes or bacteriocins are secreted into the environment, they become public goods from which all similar bacteria can benefit. As ever, there is the possibility that cheating cells can free ride on the efforts of others, undermining cooperation. The mechanism that prevents this happening in bacteria is the same one that preserves cooperation among people when they choose their companions. Human cooperators team up with each other, policing behaviour and excluding or punishing non-cooperators. Bacteria also form teams of like individuals, but they do it the microbial way, surrounding themselves with clones. Kin selection should guarantee that because clones carry the same genes they cooperate with each other, not cheat. A study of the bacteria in the human gut put this to the test.

In all, 101 different bacterial species were studied.[3] The question was: do more closely related gut bacteria cooperate with each other more, as Hamilton's rule would predict? Relatedness between bacteria in each of the 101 species was determined from gene sequences isolated from stool samples. Because bacteria divide clonally, but to varying degrees, it was no surprise that the average relatedness of two cells belonging to the same bacterial species was greater than a half—closer than siblings in a human nuclear family. But relatedness did vary across species, being nearly 1 in some species and as low at 0.1 in others. The degree of cooperativeness in each bacterial species was measured by counting the number of genes it possessed for producing public goods. Just as predicted by kin selection theory, the more closely related the members of a species were, the more cooperative they were.

Bacterial cells can move about, but most are also able to fix themselves to a surface by secreting a variety of sticky substances that create a biofilm. Biofilms form everywhere there are bacteria, from wounds on the body to the rocks in fast-flowing rivers.[4] Stromatolites are the mineralized accumulation of biofilms and mats of microbes. A biofilm formed on your teeth and inside your mouth since you brushed this morning. There could be 500 bacterial species in there. If this leaves you spitting, relax, because within the bounds of oral hygiene this is normal and some of those bacteria are keeping pathogens in check.

Biofilms protect the bacteria in them from predatory bacteria,[5] antibiotics, and the immune system, so from the perspective of individual cells, cooperating with neighbours to create a film is beneficial. The immediate neighbours in a biofilm are likely to be clonemates, but biofilms typically contain many different species. What may appear a flat film to us, close-up is more like a microcosm of Gotham with towers of cells interwoven with fluid-filled channels. Models show that such structures can be built from the simplest of interactions among neighbouring cells and do not require wide-scale cooperation.[6] But bacteria do communicate.

Animals communicate for the purposes of mating and the defence of territory, but neither of these apply to bacteria. Why do bacteria need to communicate with each other? The answer goes back to the fundamental fact that bacteria, like the famous Pompidou Centre in Paris, have their guts on the outside. Bacteria break down their food with externally secreted enzymes, they attack competitors by lacing the environment with bacteriocins, and they build biofilms by leaking glue into the neighbourhood. All these functions produce public goods that require the force of numbers for them to work best. But simply pumping out public goods without a check for the presence of other cooperators would lay a bacterium open to wasting resources and exploitation by cheats. This is when communication helps.

Using QS, bacteria signal their presence to each other, enabling them to trigger the production of public goods when there are sufficient cooperators. Recall that QS was first discovered in the bioluminescent bacterium *Vibrio fischeri*, which does not emit light until bacterial cells reach a sufficiently high density. Hawaiian bobtail squid exploit this mechanism to manipulate their symbiotic bacteria, switching them on at night and then extinguishing the light show by expelling the bacteria every morning. The function of the borrowed bioluminescence for the squid is camouflage to hide from predatory fish. Perversely, the function of bioluminescence for the bacteria themselves appears to be the very opposite: to attract animals that will eat them.

There are many species of bioluminescent bacteria in the sea. Experiments with one of them involved placing a bag of luminous bacteria at one end of a seawater aquarium and an identical bag of dark mutants of the same bacteria at the other end. The tank also contained tiny shrimp and other zooplankton. After 15 minutes, most of the zooplankton were clustered around the glowing bag, showing that they were attracted to the light coming from it, like moths to a flame. Follow-up experiments found that zooplankton that consumed luminescent bacteria became luminescent themselves, and that at night these shrimps were the ones eaten by fish.[7]

The aquarium experiments uncovered a process that we now know must also happen in the open ocean. Atlantic cod are large predatory fish, and their guts, like those of all animals including our own, contain a miniature world of bacteria and other microbes. This world, which varies with diet and between species, is known as the microbiome and much of it is symbiotic, supplying the host with nutrients. The cod microbiome is 50 per cent luminescent bacteria.[8] This discovery leads to the inescapable conclusion that the natural habitat of luminescent bacteria is not just the open sea, though they are found there, but the guts of predatory fish.

Such bacteria, which are regularly voided by fish, apparently use luminescence to find their way back into fish guts via zooplankton in the food chain that leads back to cod. Bacteria are short-lived, so a bacterium that is pooped out by a fish does not itself reach another cod, but some of its descendants will. The presence of luminescent bacteria in the gut microbiome explains why *Vibrio fischeri* is so at home inside Hawaiian bobtail squid and indeed how it could live for months in the guts of a scientist studying them. A related species, *Vibrio cholerae*, is the waterborne, intestinal pathogen that causes cholera.

Bioluminescence and the production of other public goods consumes resources. QS ensures that this costly production is only deployed when cooperators are present. But how vital to their success is it that bacteria use QS to control cooperation? What would happen to a mutant that produced public goods without checking for the presence of cooperators? In an experimental investigation of this question, evolution was compared between cultures of *Vibrio* with intact QS ability and cultures of the same bacterium containing a mutation that turned the QS switch permanently on so it produced public goods unconditionally.[9]

In both kinds of culture, within a few days non-cooperating mutants spontaneously appeared that produced little or no public goods. What happened next depended on the presence of QS. Cultures of bacteria with normally functioning QS limited the advance of the non-cooperators, but cultures of bacteria with QS always on were rapidly overtaken by non-cooperators. The conclusion from this and similar experiments is that QS enables bacteria to safely cooperate in the production of public goods of many kinds. Non-cooperators and cheats arise frequently, but QS genes coupled to the genes for public goods limit their spread.[10] Just as we saw in the human realm, communication is essential to cooperation.

QS is probably present in most bacteria, suggesting that conditional cooperation *within* bacterial species is common. Cooperation between species should be less common because one of the chief mechanisms at work among clonally dividing cells—kin selection—cannot operate in unrelated organisms. There are direct advantages to cooperating with non-kin, and we will come to those, but bacteria are well armed for hostilities.

The first weapon in every bacterium's armoury is to starve the competition by soaking up all available resources. Because bacteria can increase at an exponential rate, simply arriving at the dinner table first almost guarantees there will be nothing left for latecomers, though putting up some chemical defences against rivals also helps. The antibiotics that we use in medicine against bacteria, such a streptomycin, chloramphenicol, and tetracycline, were first discovered in soil bacteria, where they evolved as chemical agents against other bacteria.

Another class of chemical warfare agent used by bacteria against each other are the quorum quenchers. These are enzymes that destroy the signal molecules used in bacterial QS. This directly interferes with bacterial cooperation and the production of public goods.[11] A combination of a quorum quencher and an antibiotic is an especially effective weapon, since the first prevents the formation of biofilms that protect bacteria against the second.

Many bacteria possess nano-scale injection machines that enable them to stab other bacteria and eukaryote cells, injecting them with proteins and lethal toxins. One type of these nano injectors, common in pathogens such as the cholera bacterium, is virtually identical to the device that viruses use to inject their DNA into bacteria, minus the head of the virus itself. Whether this machine has been co-opted by bacteria from viruses or the other way around is not known.[12] Either is possible.

The common soil bacterium *Burkholderia thailandensis* uses its injector system to kill non-cooperators. Hundreds of genes are linked to *Burkholderia thailandensis*' QS system, including those that produce public goods and one that makes a molecule that behaves as a friend-or-foe identification signal that disarms the injector. Mutants that take public goods without reciprocating are injected with poison when they come into contact with cooperators because the mutants produce no signal to identify them as friendly.[13]

Despite the bristling armaments, bacteria live in rich communities of different species—from the metabiome of hundreds of species and trillions of cells in your gut to the bacteria that abide in soil and everywhere else. A remarkable study set out to pitch bacterial species from a soil microbiome against each other in one-on-one contests to determine what proportion would be competitive and what proportion would be cooperative.[14] A staggering 180,408 pairs of 20 bacteria growing on 40 different food sources were evaluated in a miniature robotized laboratory. Just over one-third of the interactions were competitive—defined as both bacterial species doing worse in a mixture than when growing on their own. One-fifth of interactions were parasitic, where one species did better and the other worse when the two were cultured together. In only 5 per cent of mixtures did both species do better than when growing on their own. These are the mutualistic interactions where cooperation could be occurring.

This experiment and some other, smaller-scale ones suggest that bacterial competition is far more common than bacterial cooperation,[15] but such laboratory experiments may be misleading. A significant fraction of bacteria that we can identify from their genome sequences cannot actually be grown in the lab. Even for the well-studied human gut microbiome, half the species present may fail to grow under artificial conditions.[16] A possible reason for this is that these unculturable species require substances supplied by a consortium of other bacteria which are absent in lab experiments. If this interpretation is correct, then of course cooperative bacteria are the very ones that would be missing from lab experiments because they cannot be grown.

A different approach to evaluating the nature of bacterial interactions was taken by a study that used genome sequences to work out how dependent bacterial species are on one another for public goods. The genome of a bacterium can be read like an inventory of what a cell's own genes are able to make. All cells need the same basic set of nutrients, such as the 21 amino acids that are required to build proteins. A genome sequence can therefore be checked against a list of metabolic genes to see if a cell possesses the ability to make an essential molecule for itself. If it does not, then it must be dependent on public goods and other bacteria for a supply of that molecule.

The study used data from the Earth Microbiome Project, which sequenced more than 23,000 microbiome samples from water, soil, animal guts, plant surfaces, and even the air, across every continent including Antarctica.[17] The collecting, sequencing, and analysis of these samples was an extraordinary feat of cooperation in itself and testament to the power of big data applied to small organisms. The result was an eye-opener.

Bacterial microbiomes separated quite clearly into two distinct camps: communities that were predominantly cooperative in nature and communities that were largely competitive.[18] Bacteria among the cooperators had on average 40 per cent fewer metabolic genes than bacteria among the competitors, indicating a massive sharing of public goods within cooperative communities. As one would expect from this finding, bacteria in cooperative communities were significantly more unlike each other, reflecting a greater division of labour than in competitive communities.

The cooperators and competitors were distributed differently across the habitats that were sampled. Competitor communities were mostly found free-living in environments such as soil. This is consistent with the earlier experimental results that found soil bacteria grown in pairwise mixtures often compete. Cooperators were found in a greater diversity of habitats; some were free-living, others inside animal and human guts, houses, and wastewater treatment centres. Some groups of cooperating bacteria were found to occur repeatedly across different environments. This suggests that cooperative bacteria make self-sufficient teams or consortia that are tolerant of environmental variation.

The genes that were most often missing from the genomes of cooperative bacteria were those needed to synthesize an amino acid. Different bacteria had different deficiencies, such that the amino acid lacking in one species could be supplied by another. In effect, the community behaved as a team exhibiting a marked division of labour. The bigger the team, the greater was the division of labour.

The transfer of essential resources such as amino acids and vitamins between cells is known as cross-feeding and is a common phenomenon in microbial communities.[19] It is also the basis of many symbioses, such as that between plants and mycorrhizal fungi. Just as symbioses can exist anywhere along a continuum, from parasitism (where one partner benefits at the expense of the other) to mutualism (where both benefit), so too can microbial interactions.

A common gut bacterium called *Bacteriodes ovatus* (*Bo*) makes enzymes that break down the dietary fibre molecule inulin, even though *Bo* does not use the breakdown products itself. Another bacterium, *Bacteriodes vulgatus* (*Bv*), found at high density in the human gut, lacks the enzymes required to break down inulin but feeds on the breakdown products provided as a public good by *Bo*. Recall that Charles Darwin staked the very survival of his theory on the observation that no species should carry out an action that was solely for the benefit of another species. Perhaps *Bo* didn't get the memo? Experiments show that *Bo* benefits from the presence of *Bv*, so by feeding *Bv* with inulin breakdown products *Bo* indirectly increases its own fitness.[20] Precisely how *Bv* benefits *Bo* is not known.

The circumstances of microbial life make reciprocal nutrient transfers between two bacterial species a risky business. For two species to evolve reciprocal dependence on each other, they must encounter one another frequently enough for the benefits to be reliable.[21] *Bv* and *Bo* can achieve this because both are dominant gut bacteria. Symbionts such as bean plants with root nodules, Hawaiian bobtail squid with light organs, and insects with a bacteriome all confine their bacterial partners in a compartment where they are protected and controlled. Bacteria can approach the same dependability by forming a biofilm, but in the soil, the human gut, or the open ocean there is huge bacterial diversity and a constant turnover of cells. It's tough out there and life is very short.

Bacteria are prey to an army of viruses called bacteriophages (or 'phage' for short), whose favoured method of reproduction is to convert the machinery of the bacterial cell to virus manufacture and then to explode the spent factory and spread its infectious contents. More than 200,000 different phages are known from the human gut.[22] Under the onslaught of phage, the half-life of the average bacterial cell is just two days. Phage that infect and burst marine bacteria liberate a billion tons of carbon into the ocean every day.[23]

The consequence of such mayhem is that live bacterial cells are bathed in a nutritious broth of broken ones. Add to the funeral fayre the enzymes and other resources that living bacteria leak into their environment, and you get a picture of the rich buffet of public goods available to all. In these circumstances, experiments have shown that bacteria evolve dependence on public goods for essential resources and they lose the redundant genes that enable them to make their own.[24] Does the dependence of bacteria on public goods lead to cooperation?

The answer depends on how you define cooperation. Bacterial communities in the Earth Microbiome Project that depended on sharing many public goods were described as cooperative, but dependence and cooperation are not the same thing. For example, tropical and sub-tropical seas swarm with a minute photosynthetic cyanobacterium called *Prochlorococcus*. This originally evolved in the lower-light environment of deeper waters, from which a strain of the bacterium evolved that lives in surface waters where light intensity is much higher.

Unexpectedly, *Prochlorococcus* strains from the high-light zone can't survive under intense light in lab cultures, unless non-photosynthetic bacteria are added. The problem is that intense sunlight exposes photosynthetic bacteria to high levels of hydrogen peroxide from seawater. This is a strong oxidizing agent and bacteria killer, as well as the secret of bottle blondes. Non-photosynthetic bacteria in surface waters secrete an enzyme that destroys hydrogen peroxide. They do this so efficiently that *Prochlorococcus* has lost the capacity to protect itself from hydrogen peroxide and has consequently become dependent on the rest of the community for its survival. In fact, *Prochlorococcus* from surface waters is so dependent on the protection provided by other bacteria that it has become more sensitive to hydrogen peroxide than strains of the same bacteria that live in deeper water.[25] Now, here is the poser: does this dependence represent cooperation from other bacteria?

According to the definition that we have been using, cooperation is a relationship between individuals in which one or both benefits the other at a cost to itself. This definition does not describe the relationship between *Prochlorococcus* and the other bacteria in its community, even though *Prochlorococcus* does benefit. What is missing is that the other bacteria pay no cost to help *Prochlorococcus*. They simply leak an enzyme into the water that as a by-product happens to make it more habitable for the cyanobacterium. This kind of relationship is described as commensal. How common is bacterial commensalism, compared to cooperation?

The study of the Earth Microbiome Project described many communities as cooperative, but this was done on the basis that bacterial species shared many resources, and as we have seen, sharing public goods can just as well be a sign of commensalism as cooperation. To tell which it is we need to know what the costs and benefits are between interacting species. Unfortunately, this cannot be done just using sequences, and the growth experiments we have considered, albeit with their limitations, suggest that cooperation between species in microbiomes is rare.

In the absence of more experimental information there are two reasons to expect bacterial commensalism to be more common than cooperation. The first is the turbulent, diverse nature of microbial communities that we have already discussed. The second, related to the first, is that changing the environment of a microbiome can radically alter the nature of interactions among the community.

For example, four species of bacteria found growing in the toxic environment of metalworking fluid had mostly positive effects on each other's growth because jointly they were better able to detoxify the fluid. However, if the toxicity of the fluid was reduced, if nutrients were added, or if more bacteria were introduced, the interactions changed to become more competitive.[26] Similar results have been found among soil bacteria exposed to toxic copper.[27] These are further instances of harsh environments favouring positive interactions, as previously described in breeding birds and alpine plants.

Cooperation may be the gold standard of teamwork, but a commensal team can also be very productive. The beverage kombucha, popular as a health drink, is made by inoculating cold, sweetened tea with a substance called a SCOBY, which stands for a 'symbiotic culture of bacteria and yeast'. The kombucha SCOBY is a small microbiome all to itself that contains a handful of yeast and bacterial species that form a floating gelatinous raft on the surface of the tea.[28]

Yeast cells secrete the enzyme invertase into the solution where it becomes a public good converting sucrose into glucose. Bacteria take up most of the glucose and produce a biofilm that makes the raft. The bacteria are the more dependent partner in the SCOBY because yeast can function on their own. However, the bacteria do have an indirect role in yeast-to-yeast cooperation. Bacteria can compete more effectively for glucose than cheating yeast that use the glucose but produce no invertase.[29] With bacteria thus defending the glucose productivity of cooperating yeast, the SCOBY grows with each fermentation. As a result, anyone who makes their own kombucha soon has SCOBY to spare, making it both the product of microbial public goods and a public good you can share freely with your friends.

Yeasts do love a good party and when there are no bacteria around to limit cheating, other measures are needed to police the gate crashers. In some conditions there is a natural limit on the success of cheats. In experiments with mixtures of yeast cells suspended in a liquid medium—in a beer vat for example—cheats do well when mixed in with cooperators at high density. Like a pickpocket in a crowded bar, a freeloading yeast suspended in such an environment has many sources of income.[30] But a well-mixed beer vat is not a natural yeast habitat and out in the wild they more normally inhabit plant surfaces or live in viscous pools of sugary sap. What happens to cooperators and cheats living on surfaces as they do in the wild?

When cooperators and cheats were grown on the surface of plates of agar gel, cheats had disappeared by halfway through the experiment.[31] Not only did cooperators resist invasion by cheats, but over the course of the experiment super-cooperators evolved, carrying extra copies of the gene for making invertase. In an environment where budding cells form colonies, neighbours are clonal copies of each other, or are at least highly related, and kin selection favours their cooperative behaviour towards each other. A cheat, on the other hand, finds itself surrounded by its own kind and prospers no better than a thief at a pickpocket's convention.

Any gene that has effects that increase the proportion of its copies in future generations is favoured by natural selection. Some such genes are shared between unrelated individuals—can they favour cooperation? William Hamilton suggested that if a gene for cooperation could somehow identify copies of itself among non-relatives, then it ought to direct cooperation towards them as well. Richard Dawkins imagined such a scenario: if a gene for cooperation caused the person carrying the gene to grow a unique characteristic such as a green beard, then green beards should recognize and help each other, even if they were not relatives.

Hamilton and Dawkins both considered that greenbeard genes, as they came to be known, would be most *unlikely* to exist in real life because it seemed improbable that one gene would be able to do both jobs of influencing cooperation and at the same time producing an identifiable mark that could be recognized by others carrying that same gene.[32] Decades of subsequent research have shown that Hamilton and Dawkins were half right. If you are looking for a greenbeard among animals, it had better be St Patrick's Day in an Irish pub. But if you want to find greenbeards among microbes, you need look no further than ordinary beer.

When a batch of beer has been brewed and yeast have turned the available sugar into alcohol, the yeast obligingly flocculates, forming dense clumps of cells that sink to the bottom of the brewing vessel, allowing the brewer to siphon off a clear product. Yeast differ in their propensity to flocculate, which depends on the stickiness of the cell surfaces. Several genes are involved in this variation, but one gene in particular, called *FLO1*, has a strong effect, causing cells carrying it to stick to each other. In experimental mixtures of yeast cells with or without the sticky version of *FLO1*, it is predominantly the sticky ones that flocculate and the others that get left behind in the liquid medium. So here is a gene that seems to be able to recognize the presence of copies of itself in other cells and to team up with them. Could *FLO1* be an elusive greenbeard gene?

For *FLO1* to qualify as a greenbeard, flocculation has to be beneficial to the cooperating cells, which indeed it is. Clumps of flocculated cells are more resistant to damage from poisons in the liquid environment, including alcohol and an anti-fungal agent made by bacteria. As with all cooperative traits, there is a downside or cost to flocculating, which is that cells inside clumps are so crowded that their growth rate is reduced. Yeast minimize this cost by only flocculating once conditions for growth are already compromised by the accumulation of toxins. The protection from toxins is greatest for cells in the centre of a clump and least for those on the outside. Cells without the sticky *FLO1* do latch on to clumps, but they tend to be on the outside where they are most exposed to poisons and where their presence protects the cooperators inside. So *FLO1* not only qualifies as a greenbeard gene but also manages to exploit potential freeloaders for its own protection.[33]

If we didn't know that they were mindless, we'd have to admire the ingenuity of microbes in managing their social relationships entirely by manipulating public goods. By this means they synchronize cooperation with kin, exchange with strangers, compete for resources, police cheats, and attack enemies. An MTE occurred when some of these external social functions were internalized through symbiotic cells combining.

Plate 2 Portuguese Man O'War

Plate 1 Two female burying beetles tending their larval offspring inside a carcass

Plate 5 Long-tailed tits

Plate 3 Chicks beg for food in a carrion crow nest

Plate 4 Eggs in a greater ani nest; the white one is recently laid by an interloper

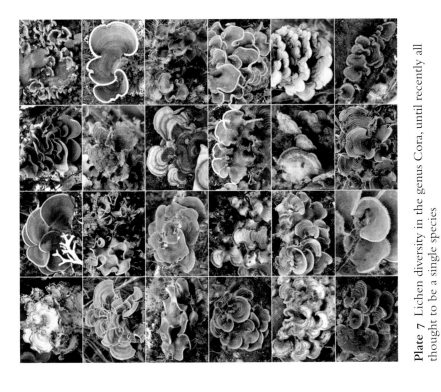

Plate 7 Lichen diversity in the genus Cora, until recently all thought to be a single species

Plate 6 Interior of a leaf-cutter ant colony

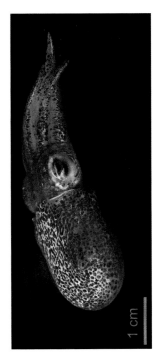

1 cm

Plate 8 Hawaiian bobtail squid

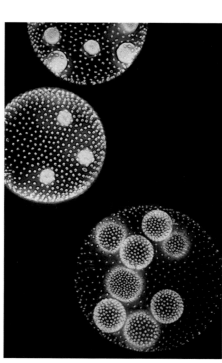

Plate 10 Volvox, a multicellular green alga

Plate 9 Grand Prismatic Spring at Yellowstone National Park. The bright rings of colour around the edge of the pool are various bacteria and archaea

10

Three-card trick

If nature is a magician, then endless cell duplication was just her theatrical debut. Once microbes got going, they began to diversify until today we are surrounded by the most stupendous menagerie. It's impossible to say how many microbial species there are because they are not only unimaginably numerous but also promiscuous with their genes and quick to adapt and evolve. Despite all this variation, a clear dichotomy has been recognized since the nineteenth century between single-celled microbes such as amoebas or yeast that possess a nucleus and bacteria that have none.[1]

Imagine this as the opening hand of a card trick in which nature has dealt two cards: the eukaryotes with a nucleus and the prokaryotes without. A card sharp will show you the faces of two cards, perhaps an ace and a queen, turn them over and swap them about in front of you. You then need to say which card is the queen. This is so easy you will confidently point to the right card. Likewise, it is easy to call any cell as either a prokaryote or a eukaryote because the difference between them is vast. This difference is as big as that between a single-celled alga and a palm tree, wrote Ernst Haeckel, known as the 'German Darwin', in 1866.

Bacterial cells are internally simple, while eukaryote cells are larger and assemble into organisms of all sizes, from single-celled amoebas and yeast to multi-celled microbiologists (Figure 15). Eukaryote cells have a complex internal structure of membrane-bound compartments. The nucleus is contained in a membranous bag that protects a eukaryote's chromosomes and selectively filters what can pass between the nucleus and the cytoplasmic fluid around it. The cytoplasm contains a semi-rigid skeleton (the cytoskeleton) made of tubes that can be rearranged to allow the cell to engulf material from its environment—a talent lacking in bacteria, which can only soak up food from their surroundings.

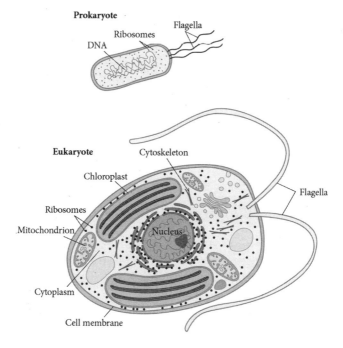

Figure 15 Eukaryote and prokaryote cells

Aside from a nucleus, the greatest distinguishing feature of a eukaryote cell are the numerous mitochondria, which by the early twentieth century were recognized, at least by some, as likely to be symbiotic in origin. With this recognition, nature was still proffering two cards (Figure 16a), though the eukaryote was now beginning to look like a composite creature, assembled like Dr Frankenstein's monster from the bolted-together body parts of at least two prokaryotes.

Then, in 1977 a scientific discovery made the front page of the *New York Times*. Carl Woese and four colleagues announced that there was a third form of life, neither ordinary bacterium nor eukaryote.[2] Nature had revealed a third card (Figure 16b). It had been there all along, hidden within the prokaryotes the way a card sharp conceals one card behind another. The reaction of the scientific community was mixed, to say the least. The dichotomy of prokaryotes and eukaryotes was so embedded in the education of every biologist since high school that it was obvious that someone must have been fooled by the cards. But was it those who saw three cards, or those who saw only two? One of Woese's co-authors later recalled that the *New York Times* article brought in a flood of phone calls, the most civil of

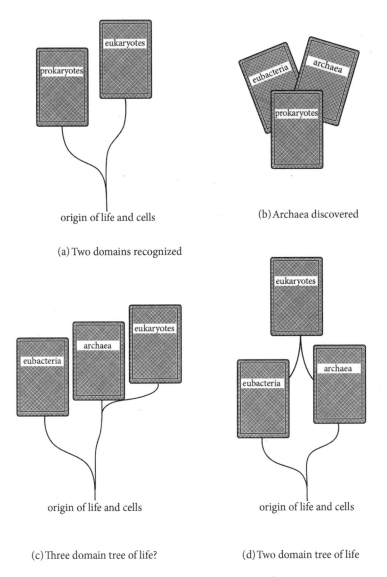

origin of life and cells

(b) Archaea discovered

(a) Two domains recognized

origin of life and cells

origin of life and cells

(c) Three domain tree of life?

(d) Two domain tree of life

Figure 16 Nature eventually shows her hand to reveal just two domains

which came from a Nobel Laureate who pleaded, 'you must dissociate yourself from this nonsense, or you're going to ruin your career!'[3] But the evidence was sound and indicated three, including a new group of prokaryotes.

These were the very early days of gene sequencing technology, but the difference detected between the two prokaryote groups was so profound that it pointed to billions of years of divergence between them. The new group of prokaryotes was very, very old. Recent studies date the divergence at nearly 4 billion years ago, making it much the oldest evolutionary event since the origin of life itself.[4]

The newly discovered group was eventually named the archaea and the rest of the prokaryotes were renamed eubacteria, though now everyone drops the formality and just calls them bacteria (Plate 9). The archaea have chemical abilities not found in most bacteria, such as turning carbon dioxide and hydrogen into the gas methane in the absence of oxygen.[5] If you poke a stick into the black, anoxic mud of a pond the archaea down there will reveal their presence with a belch of methane bubbles from the gas trapped in the sediment. As ever, these microbes do not live on their own. They obtain the carbon dioxide and hydrogen they feed on from the waste products of bacteria living in the same anoxic environment. So close are such cross-feeding relationships that a sewage sludge bacterium called *Methanobacillus omelianskii* was taken to be a single organism until it was discovered to be an association of two: a bacterium and an archaeon.[6] It should have been obvious that nothing with a name that long could be just one species.

Following the discovery of the archaea, Woese and colleagues proposed that three fundamental branches called 'domains' should be recognized in the tree of life: the bacteria, the archaea, and the eukaryotes. Now the question was: how are the three domains of life organized on the tree? Which was the bottommost and oldest branch, which the second, and which the third? Everyone could agree that bacteria had to be at the bottom as these microbes were the simplest form of life known. The last universal common ancestor (LUCA) must have been a bacterium. But what of the archaea and eukaryotes? A simple question you might think given that there are very limited ways of placing two groups on a tree, especially when one of them is related to bacteria, albeit distantly. It was not so simple though, as anyone can tell you who has fallen for the three-card trick played by a con artist in the street. Three cards are dealt out in front of you and then turned face down and very rapidly moved around. Can you find the lady? It looks easy and your confidence has been earned by success with two cards, but you can never pick the right card because sleight of hand is used to remove or hide it.

Darwin forbid that I should suggest that nature is a con artist, but who can deny that she has a wicked sense of humour? Until very recently there was almost as much uncertainty about the precise configuration of the domains as there would be if you were confronted by three cards face down. Two things were responsible for the uncertainty: the immense amount of time and evolutionary change that there has been since the events in question, and the symbiotic origin of the eukaryotic cell. Time weakens the stamp of ancient events on the genome as constant mutation blurs the picture and adaptation overwrites the record. Symbiosis and horizontal gene transfer throw a whole lot of extra genes into the mix, causing further confusion. It sometimes seems incredible that we can deduce anything at all about events four billion years ago.

Initially, it appeared that archaea and eukaryotes were sister domains, branching in different directions but at approximately the same time from their bacterial ancestors (Figure 16c). This was the three-domain tree of life, shaped like a trident with three uneven prongs. It was supported by several molecular features not found in bacteria but that were shared by both archaea and eukaryotes. An alternative hypothesis was a two-domain tree in which bacteria and archaea were sisters and eukaryotes arose later out of the archaea. In the 1990s, the evidence for the two-domain tree was weak, if only because very little was known about the archaea. Most of what was known came from the limited number of archaea that could be cultured in the laboratory.[7] These proved to be only a small and unrepresentative fraction of archaeal diversity.

The difference between a three- and a two-domain tree may seem like a distant and minor detail, but bearing in mind that we are talking about what lies at the root of the entire tree of life, this is no trivial distinction. As eukaryotes ourselves, you might say that we have a personal interest in this piece of our earliest history. It ought in principle to be possible to discover what happened by constructing an evolutionary tree of eukaryote genomes and working back down to its root. At the root of the tree, we will find the genome sequence of the last eukaryote universal ancestor (LECA). Did LECA arise parallel to archaea, making a three-domain tree, or from within archaea, making a two-domain tree?

The first stab at seeking an answer was disappointing. The earliest fossil of a eukaryote, living 1.8 billion years ago, was already a very complex cell containing all the unique eukaryote features such a nucleus and mitochondria, and leaving the same yawning gap between ancient eukaryotes and prokaryotes that we see between their modern representatives.[8] This is not an unfamiliar situation in evolutionary biology. We have two groups that we know must be related, but there is a big gap between them. Evolution proceeds in small steps and cannot leap gaps, so we have to assume that any gaps represent missing evidence—or missing links—not the sudden appearance of the eukaryotes, full -formed out of nowhere, 1.8 billion years ago.

As more and more genomes were sequenced over the first decade of the twenty-first century, the resemblance between the archaea and eukaryotes grew, but LECA was still more complex than any known ancestor and the three-domain tree remained the default hypothesis. Then in 2015 some new evidence appeared from an unlikely sounding source: mud from 3,283 metres below sea level on the Arctic Mid-Ocean Ridge, near a site known as Loki's Castle, located between Greenland and Norway. Just like soil on land, marine sediments teem with unseen prokaryotes. Though the cells are invisible to the naked eye, and most cannot be identified even under a high-powered microscope, their genomes can be sequenced en masse, and this is how we know them.

Sequencing the DNA from a small sample of mud produces a metagenome: a miscellaneous collection of genome fragments belonging to many species. Assembling the fragments in a metagenome into individual genomes is like being summoned to do the repairs in a china shop after the proverbial bull has paid a visit. But with the help of a library of reference genomes, like a set of pattern books supplied by a china manufacturer, plus some sophisticated computer programs, metagenomes give up their secrets. The metagenome from near Loki's Castle yielded sequences of hitherto unknown archaea occupying an intermediate place between prokaryotes and eukaryotes.[9]

The newly discovered microbes were clearly archaea, but they were so unusual that they represented not just a handful of new species but a whole new phylum. To a biologist, a phylum is a very compendious and highly distinctive group, such as animals with backbones (Phylum Chordata) or the land plants plus green algae (Viridiplantae). Finding a whole new phylum is therefore a big deal—like finding a new planet in the solar system. The new phylum of archaea was named the Lokiarchaeota after the site of Loki's Castle near where they were discovered. Loki is a shapeshifting deity from Norse mythology, who has been described as 'a staggeringly complex, confusing, and ambivalent figure who has been the catalyst of countless unresolved scholarly controversies'.[10] A fitting pedigree for the phylum from the deep.

The genomes of the Lokiarchaeota not only showed that they belonged bang in the middle of the domain archaea but also that they closely resembled eukaryotes. In fact, the evolutionary tree of the archaea, when reanalysed to include genomes of Lokiarchaeota and eukaryotes, revealed that these two groups were sister branches lying side-by-side within the domain archaea. The inescapable conclusion is that eukaryotes evolved from within archaea and life has two domains, not three.[11] In the final twist of the three-card-trick we are back to two cards again, but they are not labelled prokaryotes and eukaryotes any longer—they are bacteria and archaea (Figure 16d).

Eukaryotes are scions of the archaea, belonging to a group now called the Asgard archaea to which Lokiarchaeota and other recently discovered phyla also belong. Nature is having a good chuckle at the hubris of the eukaryotes who thought that they deserved a domain all to themselves just because they boasted a nucleus and some mitochondria. We have now been put in our place, and that is in the company of the Asgard.

Seeing is believing, they say, but is seeing a reconstructed genome sequence enough to justify these huge changes to the picture of our evolutionary history? There were sceptics. The news of deep-sea archaeal origins is like receiving a life-changing message in a bottle from someone claiming to be a long lost relative. The message arrives in pieces, mixed with other messages so you have to piece it together. Can such a reconstruction be accurate? The assembled pieces seem to contain all kinds of details of family history that only a genuine relative would know, but no photograph. What does this character look like? Do they have the square jaw, distinctive nose, and red hair so common in your family? Is the notorious rumour true that these ancestors dined on the waste products of bacteria?

It took a large team of Japanese researchers more than a decade to grow one of the new archaeons in culture to the point that such questions could be answered and the sceptics could be satisfied.[12] The Asgard archaeon that they cultured from deep marine sediment was a tiny cell with no mitochondria, but it was literally shapeshifting, able to produce long finger-like protrusions that often branched in the manner of a eukaryote cell. The cells were anaerobic and were found in the company of a bacterium and another archaeon that proved essential to the ability to feed on amino acids: this was a cross-feeding team. Two years later, a European team cultured a second Asgard species which likewise proved to have a bacterial partner.[13] They also obtained images of the Asgard's cytoskeleton, finally establishing the source of this feature in eukaryotes.

The discovery that archaea at the root of the eukaryote evolutionary tree form cross-feeding relationships with bacteria and seem to be capable of embracing, if not engulfing, other cells suggests several ways in which the once yawning gap between eukaryotes and prokaryotes could be bridged by symbiosis. Based on their findings, the Japanese researchers proposed that there were two symbiotic steps, each involving a major transition. In the first step, an archaeal cell teamed up with an anaerobic bacterium in a cross-feeding relationship like the ones still found in anoxic environments today. This anaerobic bacterium was then engulfed by the archaeon, becoming an endosymbiont. That bacterium has all but disappeared, leaving behind only the genes for the manufacture of its cell membrane, which became the nuclear and other compartment membranes now found in all eukaryotic cells. The second step in this hypothesis is the one proposed in 1967 by Lynn Margulis: that an aerobic bacterium joined the team, ultimately becoming the mitochondrion.

There are a number of variants of the hypothesis just described, some involving more than two bacterial endosymbionts, but one hypothesis is radically different from all the rest. In this alternative, in the first step the archaeal cell and the anaerobic bacterium fuse, but their fates are reversed, and the archaeon forms the future nucleus inside cytoplasm and membranes derived from the bacterium.[14] The second step when mitochondria are captured is the same as in the forgoing hypothesis. The archaeon-as-nucleus hypothesis sounds inside out, which it is, but it has at least one things going for it.[15] The cell membranes of archaea and bacteria are distinctly different from each other. As card-carrying Asgard archaea, you would expect eukaryote cell membranes to be like those of archaea, but they are not. They are like bacterial membranes.[16] If the archaeon ended up inside the bacterium, this would explain why the eukaryote cell has a bacterial coat on.

Genes code for proteins that do different jobs in the cell. It has been known for more than a couple of decades that eukaryote genes divide into two classes with distinct functions. A class of informational genes function in the replication of DNA and its decoding into proteins, while a class of operational genes control and code for the cellular machinery that manufactures amino acids, lipids, and other chemical building blocks.[17] The two classes of gene have different origins. The informational genes in eukaryotes can be traced back to the archaea and the deep ancestry of eukaryotes, while the operational genes come mainly from bacteria and are a more recent acquisition. For example, genes that our own cells use to detect viruses at the cell surface and which trigger the production of antiviral proteins called interferons are a bacterial legacy.[18] By one count, more than 50 per cent of eukaryote genes come from bacteria.[19] Does this mean that we *are* bacteria?

We aren't bacteria, we are eukaryotes, because the whole is more than the sum of its prokaryote parts. Eukaryote cells are a complex amalgam of genes acquired by an ancestral archaeal cell from at least two different bacterial symbionts (one the mitochondrion) and probably more. The origin of the eukaryotes is still hotly debated, but it involved at least two major transitions in evolution and all the processes of symbiotic evolution that we have seen elsewhere, from team formation to transformation.[20]

At the time of writing, the closest living relative of the mitochondrial ancestor has not been found, but we know a lot about it from circumstantial evidence. If this were a murder hunt in a TV crime drama, the evidence room would contain a wall of portraits of suspects, connected by lines of genetic evidence of who is related to whom. If you imagine that you have just joined the detective team, this is how you would be briefed.

The mitochondrion found in modern cells has been robbed of most of its genes, but there are still some present at the crime scene. From the DNA in these genes, we are reasonably certain that we are looking for a relative of the Rickettsiales, a notorious gang of bacteria including the serial killer *Rickettsia prowazekii*, which has been convicted of causing epidemic typhus in humans. Also in this group are the global parasites of insects and other arthropods known as *Wolbachia*. The modus operandi (MO) of all these bacteria is the same and like the mitochondrion itself they operate inside eukaryote cells.

The latest date at which our suspect entered eukaryote premises can be deduced from what we know about the LECA, which lived around 1.8 billion years ago. The precise location of this event is unknown, but it must have been in the ocean. Using the same forensic techniques used to identify the original archaeal ancestor of eukaryotes, investigators searched among metagenomes from oceanic mud samples and came up with some hitherto unknown members of Rickettsiales. These told an unexpected story.[21]

The mitochondrial ancestor was not a member of Rickettsiales after all but was a sister of that group, sharing a common ancestor. In another surprise, neither sister started out as an intracellular parasite, though both seem to have evolved this MO independently of each other after they parted company. One must suspect that there was either something in their shared environment or in their shared parentage that predisposed the sisters towards breaking and entering. Precisely how the mitochondrial ancestor made the transition from free-living to intracellular life we don't know, but one possibility is that it began with parasitism.[22] We have seen in other bacteria such as *Wolbachia* that there is a well-trodden path from intracellular parasitism towards endosymbiosis. Once vertical transmission of the parasite takes place, the team is on the road to an MTE. But every addition to a team must bring something to the party. What did the mitochondrial ancestor bring?

Lynn Margulis proposed that the ticket of admission for the ancestor of the mitochondrion was the ability to use oxygen in the processes that generate energy in the cell.[23] Oxygen is toxic to anaerobic prokaryotes and is a waste product of photosynthesis. As oxygen produced by photosynthesizing cyanobacteria began to accumulate in the atmosphere some two billion years ago, a symbiont that could allow its host cell to function in an oxygen-rich atmosphere would have given the eukaryotes a huge evolutionary advantage. Maybe this could explain the sudden rise of the eukaryotes after the unchallenged two-billion-year reign of the prokaryotes?

Margulis' synthesis of Earth history with eukaryote evolution was novel and appealing when it was published in 1967 because it could explain why all the then known eukaryotes were aerobic cells with mitochondria.[24] And then anaerobic eukaryotes apparently lacking mitochondria were discovered. These single-celled eukaryotes showed up in low-oxygen environments such as the guts and urogenital tracts of animals and humans. To name just two (hopefully unfamiliar) examples, *Trichomonas vaginalis* is a sexually transmitted parasite in humans and *Giardia duodenalis* is a waterborne parasite that lives in the small intestine and is a common cause of stomach upsets.

Margulis argued that these mavericks must belong to a branch of the eukaryote lineage that diverged before the mitochondrial symbiont was acquired, representing what the eukaryotes were like before they became aerobic. But further research discovered that the anaerobes after all possessed remnant endosymbionts, and to complicate matters these did not use oxygen. Like mitochondria in aerobic eukaryotes, the organelles in anaerobic ones possess their own limited set of genes—a legacy of their formerly independent existence. Could the reduced genomes of the various endosymbionts show where they came from?

Analysis of the organellar genes showed not only that aerobic and anaerobic organelles were related but convinced everyone except Margulis that all the organelles were actually mitochondria.[25] Sadly, Lynn Margulis died in 2011, but had she survived she too might have been convinced. From being the bringers of aerobic metabolism to eukaryotes, which was essential to their evolutionary success, it now appeared that the ancestors of mitochondria were more versatile than had been appreciated. They could function aerobically or anaerobically—a dual talent shared with many bacteria including the *E. coli* in your gut. Could mitochondria still be considered the gatekeepers of an oxygenated world? Yes. If anything, they now appeared even better qualified for this role. A mitochondrial ancestor that could function with or without oxygen was very well equipped to adapt to changes taking place in the atmosphere. But then another challenge to the hypothesis emerged: just how much oxygen actually was there when eukaryotes appeared?

In 1967, when Margulis published her groundbreaking paper, there was no accurate timeline for the oxygen concentration of Earth's atmosphere. It was known that there was a big uptick in concentration between 2.3 and 2.1 billion years ago because iron in rocks of this age shows the first evidence of rust (an iron oxide). This interval, now dated to an earlier 2.4 billion years ago,[26] was dubbed the Great Oxidation Event (GOE) and was attributed to a burst of oxygen produced by photosynthesis in cyanobacteria, raising the concentration of the gas to near modern levels of 21 per cent. This would certainly have provided eukaryotes carrying aerobic mitochondria with a sudden new opportunity, but was that what really happened?

Data on past atmospheric composition now indicate that two billion years ago oxygen concentration was between 0.2 and 2 per cent, far lower than today. The GOE did produce an oxidizing atmosphere, but oxygen did not rise rapidly to modern levels. On the contrary, oxygen hovered at scant concentrations until plants colonized land around 500 million years ago.[27] Thus, for the first 1.3 billion years of eukaryote history, oxygen was relatively scarce in shallow waters and absent from the deep ocean.

Even if there was little oxygen in shallow waters compared to today, there was probably enough to power aerobic eukaryotes and to poison anaerobic rivals. And of course, oxygen would have reached higher than average concentrations in environments where cyanobacteria grew best. Perhaps the trysting place that made the eukaryotes was in a mat of cyanobacteria atop a stromatolite growing in shallow, well-illuminated waters. It is a truth universally acknowledged that an archaeon in possession of good fortune must be in want of a symbiont.

Where does this leave the Margulis hypothesis for the origin of eukaryotes? The answer largely boils down to whether the bacterial ancestor of the mitochondrion was aerobic, as Margulis believed, or metabolically versatile and able to function in both aerobic and anaerobic environments.[28] Opinion is currently divided about this,[29] but there is a general consensus that the rise of oxygen in the GOE did ultimately give eukaryotes an edge, whether it was the trigger for the major transition in which mitochondria were originally acquired or not.[30] By whatever means mitochondria got to join the crew, this was just chapter one in their odyssey aboard the good ship Eukaryota.

Mitochondria are imprisoned in the cytoplasm, which means they are transmitted only down the female line when eukaryotes have sex. This narrow channel of transmission sealed their fate as subordinate partners in the symbiosis. They possess some of their ancestral genes, but the majority have been transferred to the nucleus, robbing mitochondria of their independence and sealing the major transition. Within the cell, mitochondria still divide like the bacteria that they once were, but they now depend on the assistance of genes in the nucleus to do this.

While uniparental (maternal) inheritance constrains the ability of mitochondria to defect from the eukaryote team, it's not a complete solution so long as mitochondria contain some genes of their own. These genes contain the seeds of revolt against the control of the nucleus, potentially weakening cooperation by generating mitochondrial cheats. Any maternally inherited gene that can skew the sex ratio towards females will spread as a result. This kind of selection, which is found in *Wolbachia*, also operates on mitochondria. Its effects are seen most clearly in plants, probably because plant mitochondria retain more of their own genes than do those in other eukaryotes. The place to observe the effects of this for yourself is on a grassy slope, warmed by the sun, fragrant with the scent of thyme and scattered with the wand-like flower stalks of the long-leaved plantain as they dangle their pollen-shedding anthers in the breezes of early summer.

Eye-to-eye with these lowly plants, or better still eye-to-lens, focused on their flowers, you will be able to see their sex organs. Like the majority of plants, wild thyme and long-leaved plantain normally have hermaphrodite flowers containing both anthers (the male organs) and stigmas (female). Unless, that is, there is mitonuclear conflict. This sounds like a third world war, but it's the cell biologist's way of saying that the genes in the nucleus and the genes in mitochondria have had a falling out. Conflicts are normally about power or sex. This one is about sex.

The tell-tale sign of mitonuclear conflict in the meadow is not a mushroom cloud or a ticking thyme bomb but the presence of female plants in an otherwise hermaphrodite population. In plants such as thyme and long-leaved plantain, mitochondria carry a gene that causes male sterility in the host, turning hermaphrodites into females.[31] As long as there are still a few hermaphrodites with male organs around to fertilize the seeds, male sterility can sweep through a population, and you will find an abundance of females on your field trip.[32] The vast majority of genes are in the nucleus, so when a mitochondrial gene interferes with reproduction, this is like a tiny tail wagging a very large dog, and the nuclear genes don't like it.

Half the transmission of nuclear genes is through Dad and half through Mum, so male sterility will on average halve their transmission, which is very bad news for fitness. This creates strong selection for nuclear genes that restore male sex. But it takes time for restorer genes to appear and then spread, and meanwhile females have their moment in the sun. The conflict can surge to and fro between the parties, so the sex ratio varies widely between populations. Let the smell of thyme be a reminder that cooperation is not a permanent fixture that can be taken for granted.

The ultimate move in the mitonuclear conflict is for the mitochondrion to be stripped of all functioning genes, but it appears that where this has happened, it has been at great cost. The few eukaryotes that lack a mito-chondrial genome are unable to generate energy for themselves and have become energy parasites inside the cells of other eukaryotes.[33] These para-sites made a Faustian bargain, achieving control over their cytoplasm at the expense of the independence of the cell itself.[34]

It is difficult to overstate the importance of the major transition that cre-ated the eukaryote cell, because it made complex, multicellular life possible. Almost everything living that you can actually see around you is made of eukaryote cells, including you. There is no scientific consensus yet about whether some notable features of eukaryotes, including sexual reproduction and all the internal compartments and membranes in the cell, evolved before or after the major transition.[35] It's quite possible that they were brought to the party by the archaeal ancestor, but regardless, it was the power of mitochondria that propelled eukaryotes thenceforth.

For more than a billion years the eukaryotes fuelled their mitochondria just as the ancestral prokaryotes had done—scavenging molecules from the environment to burn in the powerhouse. Then another MTE happened, creating a group of solar-powered eukaryotes that would change the world: the plants.

II

That green new thing

In his own rendition of James Brown's classic song, Otis Redding, arguably an even greater soul singer than the godfather, says that *his* new bag is green:

> Ain't you hip, of that new green thing. It ain't no drag
> Papa's got a brand-new bag.

The Soul Sister No. 1 of photosynthesis was a cyanobacterium and as we shall see later, she acquired many new bags in green, red, and brown.

The GOE around 2.4 billion years ago was the geological debut for oxygen-producing cyanobacteria, but witnesses tell us that the evolutionary debut was much earlier. The witnesses were other bacteria that had to adapt to an environment containing a rising concentration of oxygen. The genes of these bacteria indicate that enzymes enabling them to process oxygen appeared in a flurry of evolutionary innovation 3.1 billion years ago.[1]

Everything in evolution has antecedents, but they are not always what you would expect. Looking at the genes of cyanobacteria reveals two apparently contradictory things about their precursors. First, the genes for oxygen-producing photosynthesis evolved long before the cyanobacteria emerged as a group.[2] But second, the immediate ancestors of cyanobacteria were not capable of photosynthesis.[3] How can this be? At first sight it seems paradoxical, but imagine an analogous situation. No one in my family has ever owned a hat, but I'm wearing a really old vintage one. How did I get this hat? I cannot have made or inherited it, so I must have acquired it via horizontal hat transfer. Likewise, the genes for photosynthesis must have been acquired by cyanobacteria through horizontal gene transfer from a different group of bacteria.[4]

The source of cyanobacterial photosynthesis has not so far been identified. Indeed, the bacterial group that gave cyanobacteria their solar power may be extinct.[5] Those anonymous bacteria may even have succumbed to competition from the very cyanobacteria that received their photosynthesis genes. Though who knows what might turn up in the unexplored microbiome of shallow waters somewhere on the planet? Until then, cyanobacteria with their vicarious gift are the oldest known group of organisms capable of oxygen-producing photosynthesis. All other organisms with this ability, including eukaryote algae and seed plants, owe this talent to cyanobacterial endosymbionts and the chloroplasts that they became. Mitochondria derive from a single event of endosymbiosis, and likewise chloroplasts.[6] And like the transformation that made the mitochondrion, the transformation that created the chloroplast marked an MTE.

Since plants are eukaryotes, we know that the chloroplast must have been acquired after the mitochondrion which all eukaryotes have. This sets an early boundary on the date, but there was a long interval between the appearance of the eukaryotes around 1.8 billion years ago and the date of 1.25 billion years ago when it has been estimated a eukaryote cell teamed up with a cyanobacterium.[7] The identity of the eukaryote host is not known, but it must have been able to engulf bacteria, a characteristic that has been lost in most of its descendants because they have rigid cells walls. But there is a recent discovery that belongs on a branch down near the root of the green algal evolutionary tree that is a possible nearest living relative. It is a green alga called *Cymbomonas tetramitiformis* that swallows and digests bacteria during a stage of its life cycle when it lacks a rigid cell wall.[8] *Cymbomonas'* evolutionary position and ability to imbibe bacteria mean it could resemble the original eukaryote host.

Like the mitochondrion, the chloroplast has lost most of its genes to the nucleus, but it has a remnant genome and from this it is possible to identify its closest relative among living cyanobacteria. This is a species called *Gloeomargarita lithophora*.[9] The 'gloeo' part of the scientific name signifies that the bacterium surrounds itself in a gluey envelope with which it forms biofilms on rocks in fresh and brackish waters. Bacterial mats seem to be age-old meeting places for would-be symbionts and in such an environment a eukaryote may have picked up the ancestor of the chloroplast.[10]

In every evolutionary history that we have followed until now, the growing library of gene sequences has illuminated origins and clarified past events, albeit within customary limits of certainty. But the story of what the chloroplast did next, after its first encounter with the eukaryotes, has only got more complicated as the data have accumulated. What has been revealed

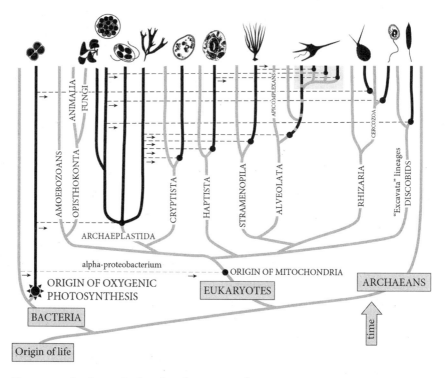

Figure 17 A schematic showing the origin of new photosynthetic lineages (black lines) and transfer (dashed lines) of symbiotic photosynthesis across the tree of life. The icons (not to scale) represent, left to right, the photosynthetic groups: cyanobacteria, land plants, green algae, glaucophytes, red algae, cryptophytes, haptophytes, brown algae and diatoms, dinoflagellates, Paulinella, chlorarachnio-phytes, and euglenids. Based on Sanders (2022). Icons from phylopic.org.

is a tapestry of relationships with a warp of vertical eukaryote lineages interwoven by a weft of horizontal transfers in which chloroplasts have been carried sideways between eukaryote hosts, often shuttled inside earlier host cells. The pattern as we currently understand it is shown in Figure 17.[11]

The initial symbiosis was so fruitful, it produced not just one but three new photosynthetic lineages of algae: reds; greens, from which land plants eventually evolved; and glaucophytes. The new organisms were single celled and hence capable of becoming endosymbionts themselves, which many later did. If you live near a rocky seashore, you will not have far to travel to see living representatives of most of the characters in this endosymbi-otic drama.

Individual cyanobacteria cannot be seen with the naked eye, but en masse they create identifiable fieldmarks in places where algae have difficulty growing. Cyanobacteria living within the surface of old seashells and porous limestone rocks stain them black. Dark green, almost black mats of cyano-bacteria form on sand bars.[12]

Though they started off as single-celled organisms, the green and red algae later evolved multicellular species. Green algae are most common in fresh water, but some grow near the top of the shore, such as the brilliantly green sea lettuce (*Ulva* spp.) and gutweed (*Blidingia* spp.) that hangs in rib-bons from vertical rocks and harbour walls. Red seaweeds come in many guises, including *nori*, which is the wrapper used in Japanese sushi. Pink species that look like branched corals or encrustations on rocks have a calci-fied armour that protects the algae against marine snails that graze every surface when the tide comes in. On the lower shore are feathery red species and in deeper water are reds that find a perch on the fronds of other, bigger algae. Below the tideline lives the red leafy sea beech (*Delesseria* spp.), form-ing an understorey beneath kelp.[13] The characteristic colour of red algae comes from a pigment that helps chlorophyll function in darker, deeper water.

The most familiar seaweeds are neither red nor green but brown. Kelps are spectacular giant brown algae that form submarine forests along rocky coasts. The egg wrack, bladder wrack, and many others are smaller brown algae that you can find growing between the tides. Brown algae are distantly related to the greens and reds and owe their chloroplasts to secondary endo-symbiosis. When still single celled, the non-photosynthetic ancestor of the brown algae swallowed a red algal cell and its valuable chloroplast passenger. The forests of kelp are the lineal descendants of that MTE, as are the dia-toms that swarm as single cells in marine and fresh waters.

The brown algae and their relatives acquired their chloroplasts second-hand. These organelles are still in their original membranous wrappers, identifying their origin as clearly as a branded shopping bag. The red algal nucleus that came in the bag has now almost disappeared, reduced to shadow of what it was. There is no doubt who is in charge: it's the host brown alga's nucleus and not the endosymbiont red's or its chloroplast.

The red algae did put it about a bit, creating four new photosynthetic lineages through secondary endosymbiosis, an even bigger score than the primary endosymbiosis that created them and their two sisters.[14] 'If you've got it, share it', would seem to be the motto of all the single-celled algae because the greens followed suit and became secondary endosymbionts in three other eukaryotes, creating three new photosynthetic lineages. One of these was a branch of marine eukaryotes called dinoflagellates whose motto was evidently, 'Come on in!'

The ancestor of the dinoflagellates was one of the four recipients of red algal endosymbiosis back in the day, but a branch of the dinoflagellates later abandoned this inheritance, only to regret it later. Three scions of the prodigal dinoflagellate lineage returned to the photosynthetic fold by variously teaming up with green, brown, and haptophyte algae. The haptophytes themselves, like the browns, are secondary endosymbionts, making the dinoflagellates owners of third-hand chloroplasts, or tertiary hosts.

If you have had trouble following the pattern of relationships woven by the shuttling of chloroplasts across the photosynthetic tree of life, don't worry. It's like trying to fathom who has done what to whom by viewing only the most recent episode of a soap opera that has been running since the Precambrian. Let me give you just a tally of the marriages: one primary endosymbiosis, seven secondary endosymbioses, and two tertiary endosymbioses. These add up to ten MTEs. If you want the score tallied in matryoshka dolls, the winner is the dinoflagellates with four.

There is, I regret to tell readers who have already lost the plot, a completely uncalled-for second series of this drama. About 120 million years ago there was a second union between a cyanobacterium and a photosynthetically virgin eukaryote. It gave birth to a creature called *Paulinella chromatophora*. Like the unexpected child of old age, this youngest member of the photosynthetic family has attracted a great deal of attention because of her youth.[15]

There are no prizes for guessing how *Paulinella* acquired an endosymbiont because it is an amoeba that lives by swallowing bacterial prey. The real puzzle is why it took so long. There was a gap of well over a billion years between the first union of a cyanobacterium and a eukaryote and the second one with *Paulinella*. There may have been others that we don't know about that went extinct in the interim, but that would only modify the question to: why did it take more than a billion years for a second endosymbiont to take, when the first was so spectacularly successful?

The expansive radiation of the chloroplast across the tree of life that followed the first photosynthetic union might itself be the explanation. Maybe the ten major transitions in evolution that spread the chloroplast far and wide filled up all the ecological niches that a later photosynthetic cell would need to establish itself. This is a difficult hypothesis to test, but the radiation of the chloroplast took hundreds of millions of years, which would surely have left plenty of time and opportunity for another primary endosymbiosis to take off and to establish. We need another explanation.

An alternative hypothesis is based on the observation that secondary and tertiary endosymbioses (at least nine in all) are much more common than primary endosymbioses (two), suggesting that the former are easier to establish than the latter. Studies of *Paulinella* indicate why this might be. In the 120 million years that *Paulinella* has been a photosynthetic organism, its chloroplast has lost 65 per cent of the genes normally found in a free-living cyanobacterium, but it still has three times the number of its own genes that a green algal chloroplast does. This clearly shows that while the *Paulinella* endosymbiont is a long way down the well-trodden road of integration with its host, the process is not yet complete.

The sense that there is unfinished business is supported by the fact that *Paulinella* is a sluggish creature that takes a whole week to double its numbers, while the doubling time of other amoebas is measured in hours.[16] Could this be a sign of relationship problems with the endosymbiont? To work efficiently, a host cell and its guest must synchronize growth and cell division, control the flow of genetic information to and from the nucleus, and manage energy use efficiently. Additionally, a photosynthetic cell such as an amoeba that can move about in its environment must adapt its behaviour to changing light conditions. In effect, the host must domesticate the endosymbiont, but it also has to adapt to its needs.

Now compare what a eukaryote secondary host of a green algal cell must do to make a success of the team. The green alga is a eukaryote, like its host, so the two already have a lot in common—they literally have chemistry. Furthermore, the green alga has already done the hard work of domesticating its chloroplast and this does not need to be repeated when a secondary endosymbiosis occurs. So, a secondary host gets a ready-domesticated puppy and the resident services of its trainer into the bargain. Is it any wonder that primary endosymbioses are so much rarer than secondary ones?[17] It's like the difference between having to prepare a meal from the raw ingredients and getting a ready-cooked meal delivered.

Ready meals of domesticated chloroplasts are widely enjoyed by eukaryotes big and small. Sea slugs steal chloroplasts from the algae that they feed on, keeping them for weeks or even months before replacing them with new ones. Flatworms hoard chloroplasts sequestered from diatoms. Stolen chloroplasts are popular with the rascally dinoflagellates, and one species found in the Antarctic relies on them entirely, keeping them for nearly three years. This species may well be on the path to reacquiring a permanent endosymbiont, having previously lost its ancestral one. It seems highly likely that further genome sequencing of dinoflagellates and their chloroplasts will uncover more cases of secondary, tertiary, and maybe even quaternary endo-symbiosis.[18] In this particular race, the dinoflagellates definitely have the form to win.[19]

Many formerly photosynthetic eukaryotes have become parasites by los-ing the function of their chloroplasts, though usually retaining a remnant of the organelle. The most extraordinary case is a group of red algae called the Florideophyceae, which contains more than 100 species that have all inde-pendently given up photosynthesis and instead have become miniaturized parasites, mainly on their closest photosynthetic relatives.[20] Time and again in this group, new species have evolved that parasitize their evolutionary parent. The parasites have all adopted a lifestyle in which they invade the host's cells, replace the host nucleus and mitochondria with their own, and adopt the host chloroplast.

What can only be described as a security flaw peculiar to Florideophyceae seems to explain why the group is a burglary hotspot. The problem is that they leave their windows open. Adjacent cells are connected by pores, and these are used by parasites to enter and spread around the host.[21] Presumably this begins as straightforward cheating by an alga's own mutant cells that spread as a cancer would do, before the cheats evolve into parasites able to move from host to host.

Cousin to the dinoflagellates is a phylum of eukaryotes called the Apicomplexa. The common ancestor of both cousins had a secondary chloroplast belonging to the red algae.[22] Half of dinoflagellate species retained or regained photosynthesis, but the apicomplexans abandoned it wholesale and went all in on parasitism. The first apicomplexan was dis-covered by the pioneer microscopist Antonie van Leeuwenhoek in the gall bladder of a rabbit. Another species appeared in the intestines of earwigs, and one you may be familiar with is the intracellular parasite *Toxoplasma gondii* that infects cats and their owners. The parasite that causes malaria is also an apicomplexan. This phylum means to be taken seriously.

Eukaryotes such as *Plasmodium falciparum* that causes malaria are tough parasites to attack with drugs. Pharmaceuticals that work against parasites by poisoning them need to be selective or they could cure the disease but kill the patient. One reason that antibiotics are so successful is that their molecular targets tend to be unique to bacteria, minimizing the side effects on the eukaryote host. There are some drugs against malarial parasites, but they are few and *Plasmodium* evolves resistance to them. But what if a target could be found in malarial parasites that is not shared with human hosts? The deep endosymbiotic history of the apicomplexans has planted just such a target in *Plasmodium*—a remnant of the chloroplast.

Although the apicomplexans dispensed with photosynthesis long ago, they and other parasites with photosynthetic ancestors tend not to lose the chloroplast in its entirety. Some small remnant of the organelle almost always remains, which suggests that it still serves a function that is important to the parasite. The remnant chloroplast in apicomplexans is known as the apicoplast. It is generously wrapped in four membranes, having been regifted twice. Since the apicoplast is bacterial in origin, it is susceptible to some antibiotics, and it could be the Achilles heel in toxoplasmosis as well as malaria.[23]

The major transition in which a cyanobacterium first teamed up with a eukaryote over a billion years ago started a chain of endosymbioses that ricocheted across the tree of life, greening the planet and ultimately oxygenating the atmosphere. The cyanobacterium, itself a borrower and not the inventor of oxygenic photosynthesis, was radically changed. Thousands of genes were transferred from the chloroplast to the eukaryote nucleus and the endosymbiont became an organelle—an organ of the cell, almost totally under the control of its host.[24]

Like its mitochondrial companion in the cytoplasm, the chloroplast is transmitted down the female line in most plants, though strangely not in conifers. In pines and their relatives, the chloroplast is carried not through the egg cell as it is in flowering plants but via pollen. In either case, uniparental inheritance preserves the eukaryote host's control over the endosymbiont.

The major transitions that first produced the eukaryotes and then turned some of those into the ancestors of all green plants and algae were unique in the history of life. Mitochondrial power unlocked new possibilities, including multicellular beings.

12

From solitude to solidarity

For the first two billion years, all life on Earth was unicellular. Even today, nearly all prokaryotes and most eukaryotes remain one celled.[1] Though unicells are in the vast majority, multicellular eukaryotes have the monopoly on large body size and complexity. A multicellular organism is no more nor less than a team of cells, subject to the same rules of cooperation as any group of individuals. Cooperation is favoured when it is to the individual advantage of cells to team up, but it must be protected by a mechanism that defeats cheats. How these general conditions apply to the particular case of the transition from one-celled organisms to many-celled ones is where it begins to get interesting.

The transition to multicellularity is different to most of the other major transitions that we have met. In the lichen symbiosis, for example, the cooperators could not be more different from each other, but the very opposite is the case in multicellularity. The cooperating cells in a multicellular organism are genetically identical. This alters how the team members behave towards each other, because sharing all their genes creates a unanimous community of interest. In symbioses and cooperative groups of unrelated individuals, the question is why teams form, but for multicellularity the problem is why such teams are not the norm.

Every unicellular organism on the planet can divide to produce 2, 4, 8 . . . cells. All that is needed to turn a unicell into a multicellular colony is for the products of cell division to stick together. This is so easily achieved that simple experiments have repeatedly shown this kind of simple multicellularity evolving in the laboratory. Unicells such as those of the alga *Chlorella* are gobbled up by one-celled eukaryotes called ciliates. When grown in culture with these diminutive predators, *Chlorella* rapidly developed multicellular clusters of 10–100 cells that were too big to be swallowed by the ciliates. Though forming a cluster protected *Chlorella* from being eaten, cells inside the cluster had limited access to nutrients. By 20 algal generations, the predominant form of *Chlorella* had shrunk to a cluster of eight cells that was big enough to avoid being eaten but small enough to allow all cells access to nutrients in the growing medium.[2]

Clearly, there are disadvantages as well as advantages when cells stick together. Some algae switch development between the unicellular and multicellular form and back again, depending on the threat from ciliates and bigger herbivores such as water fleas. Exposure to water that has contained predators can be enough to evoke the change, without the predators themselves being present. Algae can evidently taste danger.[3] The switch from unicells to multicells in algae is probably more often a behavioural response rather than an evolutionary one, but if so that behaviour would have evolved under selection from predation. Either way, whether a genetic or a behavioural response, this is a change that is simply made and easily reversed.

If the transition to multicellularity is both trivially easy and profoundly important, why are multicellular organisms unusual? We can split this question into the two familiar stages of an MET: the initial formation of a multicellular team and its transformation into a new kind of individual. The first stage is the easy one, requiring only adhesion among the daughters created by cell division. Though a sticky colony of identical cells may improve the average fitness of cells up to a point, they will quickly reach such a density that they begin to compete with one another and at that point cells may do better on their own. Something more is needed for multicellularity to be a success.

To revive a previously used metaphor, a football team of 11 goalies might stop a lot of goals, but it can't score higher than a nil-nil draw. Proper teamwork requires a division of roles and coordination among the players. Transformation from a colony of cells to a successful multicellular individual requires the same thing: differentiated cells. To achieve that is more of a challenge, and it requires a developmental programme.

The extraordinary thing about large multicellular organisms is that all their cells contain the same genes, even though the cells have many different structures and do different things. There are about 200 different cell types in the human body, but all carry the same 25,000 genes. Red blood cells are an exception, lacking a nucleus when mature. Somehow, a nerve cell that carries an electrical impulse from your spine to a muscle in your leg is made from the same genome as made the muscle fibres that respond to that impulse with a goal-scoring kick. Every cell is genetically complete. Proof

Figure 18 Dolly the sheep

of this stands in a glass case at the Roslin Institute at the University of Edinburgh.

Inside the case is Dolly the sheep (Figure 18). Dolly is a celebrity. She doesn't get out much anymore, though I did see her once at the University library, inside her case. Dolly's celebrity is due to the fact that she was the first cloned mammal. Scientists at Roslin took a cell from the mammary gland of Dolly's mother, reprogrammed it in the lab so that it behaved like a fertilized egg cell, and then implanted it into a surrogate ewe where it followed the normal course of embryonic development. Differentiated cells can only be reprogrammable in this way if they retain a complete genome—Dolly is proof that they do.

To get 200, or even two, different cell types from one genome requires a system of gene regulation that switches genes on and off as needed to form a specific kind of cell. And to get the right cells in the right place means that development must be context dependent. Cells communicate with neighbours and cells in some locations are programmed to die. The separation between the digits of the hands and feet is due to programmed cell death in the tissue connecting them during development. The genes that control positional development in fruit flies can be tricked in lab experiments into producing flies with legs on their heads or an extra pair of wings. Suddenly, multicellularity does not look nearly so simple a transition.

If the first easy step to multicellularity is simple colony formation, then we should describe the subsequent transformation involving cell differentiation as 'complex multicellularity'. Traditionally, it has been thought that the transition to complex multicellularity evolved only a handful of times: once in animals, once in fungi, and a couple of times in plants and algae. Recent research has uncovered additional cases and estimates now vary between 25 to nearly 50 independent origins of complex multicellularity.[4] Many of the previously overlooked cases are in the green algae where there could be 25 instances alone.[5] These vary in their size and degree of complexity, but at the lower end they include the charming *Volvox*, which the pioneer microscopist Antonie van Leeuwenhoek found 'a very pleasant sight'.

Volvox is a tiny sphere of clear jelly studded over its surface with green unicells, each one beating two little tail-like flagellae that propel the colony through the water like a beach-ball tumbling in the wind (Plate 10). The alga can reproduce vegetatively or sexually when it differentiates into separate male and female colonies. *Volvox* has just two kinds of cell: reproductive ones called germ cells and non-reproductive somatic cells. Such a simple, two-cell arrangement just scrapes into the realm of complex multicellularity—anything simpler would be a mere colony of identical cells—but this simplicity has a virtue in helping understand how complex multicellularity evolves.

When *Volvox* cells divide, a single regulatory gene acts as a switch that determines whether new cells will become somatic or germ cells. The trigger that flips the genetic switch one way or the other is how big the cell is. Small ones become somatic cells, bigger ones develop into germ cells, and the result is an individual that has a mixture of both types. An equivalent gene is also found in a close unicellular relative of *Volvox* called *Chlamydomonas*, where it performs a similar function. In the unicellular *Chlamydomonas* the switch is triggered by environmental stress such as an extended period of darkness. This causes the *Chlamydomonas* cell to become sexual. It appears that in the evolution of complex multicellularity in *Volvox*, the repurposing of a single regulatory gene may have been all that was needed to turn a unicell that turns sexual under stress into a multicell that converts some of its cells into germ cells.[6]

You might reasonably wonder whether so simple a system as found in green algae, involving just a regulatory gene, can explain the evolution of multicellular organisms with more cell types than two. The reason that it can is that regulatory genes control development and connect large networks of other genes. Regulatory genes are not just sensitive to their location in a developing embryo but also to the environment of the organism.

The green seaweed *Ulva* or sea lettuce can grow quite large but has just three somatic cell types. One forms the base of the seaweed called the holdfast that anchors it to rocks, another makes the blade of the frond (leaf), and a third type makes the cells that form a hollow stem connecting holdfast and frond. The stem-forming cells can also regenerate either of the other two cell types. So far so simple you might think, but attempts to grow *Ulva* in sterile cultures in the lab have found that in these conditions its spores will not differentiate into the three cell types that make the seaweed. It just sits grumpily at the bottom of the culture flask producing a spiky lump of undifferentiated cells. Something is missing.

In nature, the single-celled spores of *Ulva* and some other seaweeds only develop normally if they settle on a bacterial welcome mat. Some seaweeds require specific bacteria, and in the case of *Ulva* the welcome message on the mat is a bacterial QS signal, which when detected by a seaweed spore causes it to slow its swimming and to settle on the mat.[7] What advantage *Ulva* gains from eavesdropping on bacterial signals in this way is not known. It might just be a way of selecting a substrate that is firm enough to support a multicellular alga.

The division of labour among cell types that arises in complex multicellularity accounts for the advantage of teaming up, but there is an inherent danger in building an organism that contains trillions of cells: some of them will mutate in the process. Some, perhaps most, mutant cells will malfunction harmlessly, but all start with a complete genome. This means that even if rare, a mutant cell that successfully escapes its built-in mission control may be able to initiate a cell lineage of cheats, threatening the cooperation among cells that makes multicellularity work. There are many brakes on this happening, but one is fundamental to all complex multicellular organisms and is found in the life cycle of every one of them: a one-cell stage.[8]

The one-cell stage in the human life cycle is the fertilized egg from which each individual develops. One might naively think that having constructed a marvellously functioning multicellular organism of 30 trillion cells in 200 flavours that can read and write, the last thing that evolution would preserve would be an illiterate unicell, but it does, and for a very good reason. Passage through a unicell every generation ensures that all the cells in the next generation will be, at least to begin with, genetically identical, which keeps them cooperative. What would happen if evolution tried to build a complex multicellular organism that did not work this way? As it happens, there is an amoeba that can show us the answer.

The words 'slime' and 'mould' suggest somewhere dank and dark where most of us would rather not venture, but in the dank and dark of the soil reside the fascinating microscopic amoebae called cellular slime moulds. Their behaviour puts them right on the boundary between unicellularity and complex multicellularity as they skip cheekily back and forth across an MTE like they were playing hopscotch.

The feeding amoeba of *Dictyostelium discoideum* is a solitary cell that ignores others of its kind, spending its time engulfing bacteria, but when food grows scarce it becomes sociable and teams up. A starved amoeba emits a chemical signal in a pulse at approximately six-minute intervals. Other amoebae that detect this signal respond by moving towards its source and also begin to emit pulses of the same signalling molecule. The signal from the chorusing amoebae ripples outwards in waves, while phalanxes of cells begin to form as they advance towards the centre. Soon, tens of thousands of converging amoebae form conga lines of cells stuck nose to tail. Arriving cells slip into the lines from the sides, swelling them further as they converge on the centre.

With tens or hundreds of thousands of cells all arriving at the same point, the inevitable happens and they begin to pile up, but this is not so much a train wreck as an exuberant gathering of cells that adhere to each other and begin to organize themselves, from solitude to solidarity. A mound forms (Figure 19), this rises and develops a tip from which cells in the apex begin to send out signals that direct the behaviour of cells lower down. As a result, the tip becomes a finger and then one of two things can happen. The finger can fall over, becoming a slug, or it becomes a squatter version of itself, resembling a Mexican hat that elongates into a stalked structure.

A slug 1 mm in length contains an estimated half a million amoebae.[9] The slug is not just sluggy by appearance but sluggy by nature, as the whole structure, formed of hundreds of thousands of cells that were only recently autonomous, now act as one organism. The slug, buried somewhere in the soil, begins to migrate upwards towards the surface, attracted by the faintest light and following the slightest gradient in temperature. By day the slug

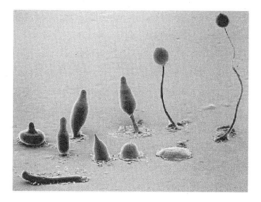

Figure 19 Micrograph of the stages in the multicellular development of *Dictyostelium*, starting with a mound of cells (right foreground) that either forms a slug (left foreground) or a Mexican hat that develops into a stalked structure terminating in a spore capsule (right). The slug is approximately 1mm long.

senses the warmth coming from the sunlit surface and moves up the thermal gradient, which must be detected across the tiny space of its body. At night, the soil is still warm but the surface cools down so the thermal gradient is reversed, but so too is the reaction of the slug, which continues to move upwards, guided now by the cooling gradient.

Once suitably positioned near the surface, the slug begins to transform itself into a stalked fruiting body that finally stands upright upon a disc at its base (Figure 19). The stalk is made from the one-fifth of the cells that were at the front of the slug. These form an internal armature of cellulose as the cells die, sacrificing themselves to provide support for the remaining 80 per cent of cells that migrate upwards and transform themselves into spores, ready to be dispersed.

Here is a social system stripped down to the barest essentials and yet capable of organizing itself to escape from conditions of food scarcity by recruiting the cooperation of a massive army of autonomous beings, 20 per cent of whom sacrifice themselves in order to hoist the majority into sunlit liberty. Two fundamental evolutionary questions arise from this. Why do cells team up? And why do 20 per cent of the cells sacrifice themselves?

The first question, 'Why team up?', is the easiest to answer. Only a team can construct the multicellular tower and fruiting body required to enable spores to be dispersed away from areas that have been depleted of food. In laboratory conditions, amoebae can be drawn from 10 mm away to join the slug. For an amoeba that is only about 10 μm (10 thousandths of a mm)

long,[10] 10 mm is a thousand times its length, or equivalent to a human run-
ner covering 18 km. From the fruiting body, spores are moved away by soil
invertebrates, such as the tiny nematode worms that occur in all soils.
Nematodes eat amoebae and bacteria, but amoeba spores can survive passage through the gut of a nematode, and as a consequence these can be
dispersed five or six times the distance over which amoebae gather to form
the fruiting body, or 100 km, scaled to human terms.[11] The advantage is
clear, at least for the 80 per cent of cells that become spores, but what about
the other 20 per cent that end up dead in the stalk?

Is kin selection responsible for this act of self-sacrifice? Are the half a million amoebae that answer the clarion call to gather just a big family, perhaps
even genetically identical? Or are stalk cells perhaps captive galley slaves,
forced to do the bidding of the victors in a cellular war? In the wild, the
great majority of cells in a *Dictyostelium discoideum* slug are indeed identical,
which means that stalk cells are very much family, more like worker bees
than galley slaves. This similarity explains their cooperation and sacrifice.
Laboratory cultures of *Dictyostelium discoideum* are different though, and in
these a type of cell often evolves that has never been found in natural populations of the species: the social parasite.[12] These mutant amoebae are unable
to make stalk cells at all, but when mixed with amoebae that can make
stalks, they can take advantage of the host stalk and use it to distribute themselves as spores. Why are such cheats, which have a 20 per cent advantage
over normal cells that sacrifice 20 per cent of their number, not found
in nature?

Lab experiments with mixtures of mutants and non-mutants provide the
answer. Cheats only prosper when thoroughly mixed with cells that are
unrelated to one another. However, this is not how amoebae are found in
nature. Amoebae, much like bacteria, multiply rapidly by cell division, which
means that they are found in patches of cells that are identical clones. In
nature, then, a dividing mutant will soon find itself surrounded by copies of
itself and, since these don't make stalks, they cannot become host to social
parasites.[13] Cheats are hoist by their own petard, rather than by a host's stalk.
These experiments illustrate an important general principle that we have
seen before, which is that high relatedness is a good defence against cheating
because cheats only prosper when able to exploit individuals unlike themselves. *Dictyostelium* does have a weak point in its cooperative defences, though.

There are two ways of forming a multicellular organism: cells can come
together (aggregate), which is how *Dictyostelium* does it, or cells can divide

and stay together, which is what happens in animals, plants, and fungi.[14] In the latter cases, the life cycle passes through a single-cell bottleneck, after which the multicellular body develops through cell division. Though the cells produced this way may specialize in different roles, they are genetically identical, which preserves their cooperation. In contrast, cellular slime moulds have no such mechanism because they produce the multicellular stage by cell aggregation rather than cell division. Forming a multicellular body this way is risky because cheats picking up the signal to congregate can gatecrash the party.

Success for a multicellular organism depends on cells cooperating for the collective good, even if that requires programmed cell death such as in stalk cells. Gatecrashers have no interest in the collective good, only in self-replication, which undermines the cooperation needed for multicellularity. Aggregation makes *Dictyostelium* a poor model for the evolution of complex multicellularity but potentially a useful one for its biggest threat: cancer.

Cancer is a disease that unwinds the cooperation on which complex multicellularity is built. It has many forms, but all involve the uncontrolled proliferation of cells—a process that at its worst can quickly overwhelm the individual. A cancer cell is a cheat bearing genetic mutations that enable it to throw off all the safeguards that evolution has built into the genome of multicellular organisms to preserve cooperation among cells. It occurs not just in animals but also in land plants, algae, and even fungi.[15] In humans, half of us will develop cancer, though thankfully it is often treatable if diagnosed early.

Mutation is a random process, subject to a throw of the dice every time that cells divide. Though mutation is rare, multicellular organisms roll the dice with the dedication of a gambling addict. Every week, all the cells in the surface lining of your gut are replaced by cell division—twice.[16] Other kinds of cell are replaced at a lower rate, but cell division is essential to body maintenance. Multiply the number of cell divisions per week by the 4,000-plus weeks in a human lifetime and even a rare event becomes almost inevitable.

Genetic studies of cancer cells show that they carry a variety of mutations in genes whose job is to supress tumours. Some of these same genes regulate the growth of cells in the multicellular development of *Dictyostelium discoideum*.[17] A common mutation in cancer cells occurs in a gene called PTEN, but this gene is hard to study in animals because it is vital to embryonic development. This prohibits determining the function of PTEN by deleting or altering the gene in an animal model of cancer. Fortunately, the equivalent gene can be found in *Dictyostelium*. Deleting PTEN in slime mould cells completely abolishes multicellularity, but *Dictyostelium* cells can live in the unicellular state. Replacing the deleted amoeba version of PTEN with a copy of the human gene restores multicellularity in *Dictyostelium*. About 1.4 billion years of evolution separate humans and slime moulds, but the function of this gene that is key to multicellularity and to its breakdown in cancer has been preserved by evolution. That is quite some endorsement of its fundamental importance.

Tracing the evolutionary histories of the 3,000 genes associated with different human cancers is illuminating. They fall into two main groups representing two bursts of evolution. The first is a set of very ancient caretaker genes that perform functions essential to all cellular life and which can be traced back to the origin of the cell. The second group is a set of more recent gatekeeper genes. The gatekeepers are involved in signalling between cells and in regulating growth. Tellingly, these originated with animal multicellularity. The PTEN gene present in both *Dictyostelium* and humans is an older gatekeeper that dates back to the origin of the eukaryotes, to which both cellular slime moulds and animals belong.[18]

Cancer cells behave like unicellular organisms to an uncanny degree, some even depending on energy generated by fermentation as bacteria and yeast do.[19] Multicellular organisms have to contend with this. When normal epithelial cells are shed from the lining of the gut and elsewhere, they are programmed to self-destruct. This cooperative act protects the multicellular organism from unregulated proliferation. This system of cancer control has a failsafe mechanism, such that its default state is self-destruction. Only the presence of neighbours prevents normal cells from dying.

Cancer cells are deaf to the proximity signals of their neighbours that normally regulate cell division and programmed cell death. In more advanced cancer this leads to metastasis in which tumour cells get distributed around the body.[20] Cancer cells have elevated mutation rates. Once a tumour has started to grow, its cells evolve as any microbe would. Just as bacteria evolve resistance to antibiotics, tumour cells can evolve resistance to anti-cancer drugs. The immune system, a defender against cancer as well as microbes, also exerts a selection pressure on tumour cells, but those cells' high mutation rate means that they are constantly throwing up new mutants that might resist the body's defences.

With rapid cell proliferation and high rates of mutation, tumours evolve into teams of rogue cells, even recruiting normal cells to their aid under a false flag of cooperation. A recent discovery is that tumours may also harbour bacteria that help them defeat the patient's immune system and inactivate anti-cancer drugs.[21] It is difficult to resist the parallel with pirates who roam at will, living outside the law while cooperating among themselves.

The sophisticated bodies built by complex multicellularity must surely count as the towering achievement of cooperation. But remember that history is written by the victors, or in this case only the multicellular can write history. Earlier in this chapter I asked, if the transition to multicellularity is both trivially easy and profoundly important, why aren't most organisms multicellular? The answers are several. First, unicells such as bacteria can achieve some of benefits of multicellularity through cooperation in biofilms and QS.[22] The pinnacle of prokaryote social achievement is found in the myxobacteria that have evolved a lifestyle uncannily similar to that of *Dictyostelium*, a eukaryote. Myxobacteria prey on other bacteria, hunting in packs that drench their victims with enzymes from their collective secretions, to absorb the liquid remains. These are predators that hunt with soup spoons. When the soup supply dries up, myxobacteria aggregate to form a spore-producing organ on a stalk, just as *Dictyostelium* does.[23]

The complexity found in myxobacteria is impressive, but only by the standards of bacteria. Apologies if that sounds patronizing, but why can't a prokaryote be more like a eukaryote? The answer may be that prokaryotes lack the necessary energy supply that mitochondria uniquely give eukaryotes.[24] But lack of mitochondria cannot be the only barrier to complex multicellularity because most eukaryotes are in fact unicellular. A simple reason for shunning multicellularity is that there are a very large number of small niches in nature successfully occupied by unicells that cannot be filled by larger organisms. There is another reason, too.

As microbes and their cancer-cell-mimics demonstrate, rapid cell division and fast evolution represent a permanent evolutionary advantage for the unicellular state. Don't take the word of one aged, multicellular eukaryote for this—consider the evidence of the single-celled yeasts that have evolved perhaps 12 times independently from multicellular fungi. If you find yourself scratching your head over why an organism should forgo the fruits of multicellularity, consider the itch a dispersal strategy of the yeast that causes dandruff. Multicellularity has been described as a 'minor major transition' because it arises so easily.[25] No one could say that of the next major transitions: the origin of genes and of life itself.

PART IV

Genes

13

Ordering the primordial soup

On 16 August 1858, Queen Victoria texted President Buchanan of the United States. The 98 words of the message took 16.5 hours to transmit and inaugurated the first transatlantic submarine telegraph service. The queen fervently hoped that the new electric cable would prove an additional link between the nations, 'whose friendship is founded upon their common interest and reciprocal esteem'. The president replied that the telegraph, produced by cooperation between the nations, was a 'triumph more glorious . . . than was ever won by conqueror on the field of battle', and wished that 'all nations of Christendom spontaneously unite in the declaration that it shall be forever neutral' so that messages would not be interfered with even in times of war. Many elements of cooperation can be read in this exchange: common interest, reciprocity, trust, the superiority of cooperation over conflict, and the integrity of communication in times of conflict. The telegraph stirred something of even deeper significance than these messages, or so thought Thomas Henry Huxley.

Two thousand fathoms deep, the submarine cable traversed the North Atlantic over a region that the marine surveyors called 'Telegraph Plateau'. Huxley was sent a sample of the mud dredged from the seabed by the surveyors. He was yet to earn the soubriquet 'Darwin's Bulldog', but he had read an advance copy of Charles Darwin's *Origin of Species* and told his friend that he stood ready to defend him tenaciously against the inevitable attacks that publication would bring. One obvious line of attack would be, 'If God did not create life, how did life begin?'

To answer this question, we need to define 'living'. In his characteristic manner, J. B. S. Haldane confessed in his book *What Is Life?*, 'I am not going to answer this question. In fact, I doubt it will ever be possible to give a full answer.'[1] Fortunately, that was not the last word on the subject, and with due deference to the complexity of life we might adopt a working definition that is frequently given: life is a self-sustaining chemical system that is capable of Darwinian evolution.[2]

Darwin himself did not address the question of the origin of life in *The Origin of Species*, wisely believing it to be beyond the power of science to answer at that time. Only much later, in a private letter to his friend the botanist Joseph Hooker, dared he speculate that life might have begun in some 'warm little pond with all sorts of ammonia and phosphoric salts,— light, heat, electricity &c. present'. Others were less coy about the question. Ernst Haeckel thought that first life would be an *Urschleim* or primordial slime that might be discovered in the ocean. Then, in 1868, Huxley went back to his jars of mud from the Telegraph Plateau that had been sitting in preservative alcohol for a decade. His report of what he found was startling, though presented cautiously and without fanfare.[3]

Huxley believed that he had found something that he had overlooked before. It was Haeckel's *Urschleim*. The mud contained an amorphous substance that under Huxley's microscope looked like naked protoplasm. He named it *Bathybius haeckelii* in honour of the German zoologist who had predicted its existence. But was *Bathybius* a living organism or just a chemical residue? Haeckel was understandably delighted by the discovery and embraced the idea that it was *Urschleim*, but others thought it might not be living at all. One critic even accused Huxley of out-right fraud.

The Scottish zoologist Charles Wyville Thomson examined some samples, noted the absence of structure, and described it as a 'diffused formless protoplasm'. Wasn't this exactly what you would expect *Urschleim* to be like—an indeterminate substance hovering somewhere between the living and the non-living? Wyville Thomson was deeply interested in what lay at the bottom of the ocean. Ships charting routes for telegraph cables were routinely sampling the sea floor for the first time in history. A previously inaccessible realm covering a large fraction of the planet was suddenly available for study and there were some vital scientific questions to be answered. For a start, how much life, if any, was down there?

Edward Forbes, professor of natural history at the University of Edinburgh, had conjectured that life could not exist at great depths. In his youth, Forbes had abandoned his early studies in medicine to work instead on the more exciting pursuit of natural history. Dredging shallow seas around the Shetland Isles was a zoologist's delight (Figure 20), which he celebrated in a 'Song of the Dredge':

Figure 20 Dredging, as enjoyed by the young Edward Forbes

> Down in the deep, where the mermen sleep,
> Our gallant dredge is sinking;
> Each finny shape in a precious scrape
> Will find itself in a twinkling!
> They may twirl and twist, and writhe as they wist,
> And break themselves into sections,
> But up they all, at the dredge's call,
> Must come to fill collections.

Samples taken in deeper seas in the Mediterranean suggested to Forbes that life dwindled away in deeper water.[4] Wyville Thomson, who in 1870 himself became professor of natural history at Edinburgh, saw an opportunity. Mustering support from other scientists, a ship, crew, and funds were obtained from the British government to mount a circumnavigation of the globe, with the dual objectives of mapping the sea floor and dredging the deepest depths to discover what life might be found there. Huxley was enthusiastic for the expedition, confidently expecting it to bring to the surface 'zoological antiquities' representative of species now extinct nearer the surface.[5] In this he was to be disappointed, but though the expedition sailed more than 150 years ago, its discoveries are still of global significance. Some even say that the expedition marked the beginnings of modern oceanography.

The Royal Navy provided Wyville Thomson and his colleagues with a refitted war ship called HMS *Challenger* for the expedition (Figure 21).[6]

H.M.S. CHALLENGER UNDER SAIL, 1874.

Figure 21 HMS *Challenger*

The ship was a hybrid, using sail for its main motive power and a steam engine for manoeuvres, including the power needed to remain on station while sampling. All but two of the *Challenger's* 20 redundant guns were removed to make room for laboratories, the team of six scientists on board, and storage for samples and a huge quantity of rope. Miles of rope were needed to lower dredges, nets, and scientific equipment to the ocean floor. The rope often broke and regularly needed to be replaced.

Wherever *Challenger* docked, samples of marine deposits and specimens of animals and plants preserved in carefully packed jars were dispatched back to Edinburgh where a dedicated office was set up to receive and store 100,000 samples. From there, material was sent to experts all over Europe and North America for study. Wyville Thomson rejected the call made from some quarters that British scientists should have first pick of the haul, insisting that the best experts should be involved, wherever they came from. When published, the *Challenger* expedition's results filled 50 large, heavily illustrated volumes authored by 75 scientists. This was big science, long before the modern term for such an ambitious enterprise had been coined.

Challenger discovered that the ocean bed beyond the continental shelf was not the featureless plain that many had imagined, but that it had a topography with abyssal depths and mountain peaks like the Mid-Atlantic ridge that the expedition mapped. Animal life was to be found at all depths of the ocean and the fauna in the deep abysses was similar across the globe.

Five thousand new species were discovered by the *Challenger*'s survey, but it was a negative finding that had greatest impact for those who believed that *Bathybius haeckelii* might prove to be primordial life. The expedition's on-board chemist, John Young Buchanan, showed that the substance that Huxley had taken to be *Urschleim* was in fact just a chemical precipitate produced when alcohol was added to seawater.

Huxley took the embarrassing news well when informed by Wyville Thomson and he issued a retraction. It took Haeckel longer to accept that his eponymous primordial slime was a fiction. Wyville Thomson had more bad news. On his return, he wrote that the expedition had found that '[t]he character of the abyssal fauna refuses to give the least support to the theory which refers the evolution of species to extreme variation guided only by natural selection'. Darwin himself replied, not mincing any words, in a letter to the journal *Nature*: 'This is a standard of criticism not uncommonly reached by theologians and metaphysicians, when they write on scientific subjects, but is something new as coming from a naturalist.'[7]

Calling another scientist unscientific like this is about as bad it gets in normally polite scientific circles. Darwin, who was usually quite retiring, was for once his own bulldog, but he had a good point to make. He argued that you could not just catalogue the results of evolution—in this case the diversity of life on the ocean floor—and expect to see the underlying mechanism of natural selection that generated it ticking away in front of you. You wouldn't expect this on land, so why expect it in the ocean's abyss?

What evolutionists desperately needed was a mechanism that would bridge the gap between the non-living and the living without recourse to miracles or large amounts of rope. In his private letter to Hooker, already mentioned, Darwin doubted that it would be possible to witness the creation of life anew in any modern warm little pond. He feared that the descendants of life generated in such places aeons ago would inevitably be present now and would consume all the ingredients before anything new could emerge. This is an excellent point and probably explains why there was just one origin of life in the first place. However, it is a testable hypothesis. Could life be experimentally recreated in sterile laboratory conditions?

If you have had the merest sniff of a biology textbook in the last 50 years, you will have read about the laboratory experiments of Stanley Miller and Harold Urey in the 1950s when they generated a variety of amino acids and other building blocks of life by passing an electric spark through a flask of gases that mimicked the early atmosphere of Earth. This confirmed the earlier ideas of the Russian chemist A. I. Oparin and of J. B. S. Haldane that the primordial ocean may have contained a soup of organic molecules generated by lightning. The ocean once again! The imagination of scientists searching for the origin of life perpetually dwells on the ocean because the fossil record is quite clear that this is where life emerged.

We will come back to the ocean as the birthplace of life, but our modern understanding of cells has opened up an approach to the origin of life that was just not available in the 1850s, or even in the 1950s. Instead of looking into the vastness of the ocean and trying to drag evidence of primeval life from the abyss, we can look into the cell itself for evidence and work backwards. Even without knowing how life began, we can deduce what barriers had to be overcome for it to turn out the way it has. This approach has led to remarkable progress in what once seemed to be an insoluble mystery.

Starting with a modern cell, the first problem is that cells depend on two quite different kinds of molecule working hand-in-glove. One kind of molecule stores and replicates the genetic blueprint of the cell, and another kind entirely is used to build and power it. The first molecule is DNA, the stuff of genes, and the second is protein, the substance of enzymes and much else. Both molecules are polymers—that is, chains of smaller units linked together, like beads on a string—but they are built of different subunits. DNA molecules are made of nucleotides, protein molecules of amino acids.

We don't need to delve any deeper into the mind-boggling structure that is a functioning cell—the leading textbook on this subject is more than 1,400 pages long—to appreciate the problem.[8] Any conceivable primitive cell would have needed molecules that performed both the functions of DNA and proteins: inheritance is essential to life, and so is the harnessing of energy, for which enzymes usually made of proteins are required. Could both functions have been performed by a single kind of molecule instead of two? And if so, how did the functions become separated as they are in all cells today?

Evolution is conservative, making new devices by tinkering with old ones. If the cell were a factory open for the inspection of visitors, you would be impressed by the frantic industry of what is going on inside, by the heat generated from thousands of mitochondria running at 50°C, the ceaseless whizzing about of transporters carrying raw materials and products, but nothing would look shiny or new. What might surprise the visitor is just how much is being done with ancient machines that have been patched and jerry-rigged ever since the factory first opened for business in a wee shed.

The oldest machines of all are ribosomes that have been running since the year dot. Ribosomes are fully functioning, four-billion-year-old molecular fossils. What can such ancient molecular fossils do? Everything.

Ribosomes translate the genetic message encoded in DNA into proteins and there are millions of them in every cell. The genetic information required to make a protein is transmitted between the DNA blueprint in the nucleus and the ribosome in the cytoplasm in a molecule called RNA. RNA, like DNA, is a chain made of nucleotides, but the RNA molecule is single stranded, unlike DNA, which is famously a helix of two strands. DNA is the chemically stable repository of genetic information, while RNA translates the information into action. Being single leaves RNA more open to relationships with other molecules, which is essential to its many functions.

Different kinds of RNA molecule do different jobs. Messenger or mRNA carries information from the nucleus to the ribosome. The ribosome itself is made of rRNA (r for ribosomal) and some protein. We know that the ribosome must be really ancient because the genes that code for its core functions are practically universal. Over billions of years, life has diversified, and most of the underlying genes have changed, but there is a core set of fewer than 100 genes that is found in all species, from prokaryotes to pro-footballers.[9] Most of those 100 genes code for rRNA. This implies two things: the core of ribosome structure arose very early in the history of life, before the last common ancestor of all living organisms, and its function is so important that it has been preserved intact by evolution.

A ribosome begins to make a protein somewhat as a novice potter starts a pot by coiling a long, thin strand of clay upon itself. In the ribosome, the process of construction begins with the strand of an mRNA molecule bearing an encoded recipe for a protein. This is fed through the ribosome like a tape. Unlike a strand of clay in a coil pot, the mRNA does not itself become part of the growing structure, but it serves as a template on which the protein is built. As the mRNA tape passes through the ribosome, amino acid molecules are delivered to it, one at a time, attached to a small transfer or tRNA molecule. There are 20 different kinds of amino acids found in proteins and they have to be assembled by the ribosome in a specific order to obtain a functional protein molecule at the end. This is accomplished by the tRNAs, which are adaptors.

tRNA molecules carry the code for a specific amino acid at one end and the amino acid itself attached at the other. Inside the ribosome, the tRNA attaches to the matching code on the mRNA tape, bringing the amino acid it is carrying into the correct position on the end of the unfinished protein. As each new amino acid is lined up, the ribosome attaches it to the protein it is making and the tRNA detaches: job done.

As assembly proceeds, the growing protein strand spools out of the ribosome's exit tunnel and spontaneously folds into a three-dimensional (3D) shape. The exit tunnel belongs to the highly conserved core of the ribosome. Every protein in every cell on the planet for the last four billion years has issued into the world this way. The ribosome exit tunnel is nothing less than the molecular birth canal of all life.

It has doubtless occurred to you by now what a good candidate the ribosome is for an early player in the origin of life. Not only is it suitably ancient, but RNA fits the description on the 'Wanted' poster for a molecule that can both transmit hereditary information and harness energy. RNA can do the first by virtue of its structure and the second through the ability of the ribosome, an RNA machine, to manufacture enzymes. To cap it all, the proteins made by ribosomes include enzymes called RNA polymerases that are used to make ribosomes. The ribosome makes the tools for its own construction. Is it case closed?

If a living fossil as old as the ribosome was found at the bottom of the ocean, it would certainly make headline news. Of course, ribosomes are indeed to be found at the bottom of the ocean in all the cells there, but they are also in our fingertips. Can anything in biology be more exciting than the idea that we are personally so close to the origin of life? Before we get too excited though, there is another obstacle to be overcome on our journey in reverse gear to the big beginning.

Versatile though they are, ribosomes depend on many other cellular components for their function. In yeast for example, some 200 different molecules are involved in the assembly of a new ribosome and the blueprint for this is held in DNA in the nucleus, not in the ribosome itself.[10] So the question is, could ribosomes have evolved from ancestors in a pre-cellular world—in a primordial soup? This is just what the RNA World hypothesis suggests happened. It proposes that RNA was a crucial ingredient of the primordial soup and the starting point for the evolution of a living cell from non-living components. Four different scientists independently hit upon the same idea in the 1960s, of whom the first was J. B. S. Haldane in 1964, a year before his death.[11]

For the hypothesis to be viable, three things would be necessary just for starters. First, the nucleotides that make RNA molecules would need to be present in a pre-biotic, primeval world devoid of life; second, these must spontaneously link together to make RNA; and finally, RNA molecules must be able to make copies of themselves. Sceptics have challenged each of these steps as very unlikely for a variety of reasons, but against the odds, experimental evidence demonstrating the possibility of each one has gradually accumulated.

Beginning with the nucleotides, these are more complex than the amino acids that Miller and Urey generated in their laboratory experiments, but a network of chemical reactions has been discovered that produces some of them in a laboratory simulation of pre-biotic conditions.[12] These reactions produce the nucleotide building blocks for DNA as well as RNA, suggesting that both kinds of polymer could have existed in the primordial soup. This is an important breakthrough in itself because if the primordial soup contained both DNA and RNA, the first genetic molecules may have been hybrids of both. It is easier to see how the molecular division of labour between DNA and RNA that exists now could have come about if the precursor was a hybrid of the two, rather than if it was just RNA.[13]

A potential problem for the hypothesis is that the chemical reaction network produces nucleotides with the correct structure for RNA and DNA formation mixed in with other, badly formed nucleotides that could poison the soup. What looks at first like a problem is elegantly resolved by exposing the reaction mixture to ultra-violet (UV) light, which eliminates the problematic nucleotides. Today, the surface of the Earth is protected from harmful UV radiation by the ozone (O_3) layer, but back at the dawn of life there can have been no ozone layer because there was no oxygen (O_2) in the atmosphere to supply it.

Could the handful of nucleotides that make up RNA and DNA have been selected from the many that could theoretically exist simply because they were unusually stable under UV radiation at the dawn of life? It seems so, in which case this fundamental piece of life's chemistry carries the hallmark of its most ancient origins. We shall see later that a requirement for UV imposes a constraint on where on the planet life may have started, assuming that these lab experiments have correctly identified the chemical route that it took.[14]

The same network of reactions that makes nucleotides also produces some amino acids—the building blocks of proteins—and lipids, which are a major component of cell membranes. With all these different building blocks of life floating about, the experimental soup begins to look promisingly primordial. But what are the chances that nucleotides in an alphabet soup will spontaneously join up and that this will spell out an RNA message that can copy itself? Once again, the sceptics would say 'no chance', and here too a riposte is found in the remarkable chemistry of RNA.

Chemical reactions can be made to happen faster and more easily, with less or even no input of energy, by means of a catalyst. Enzymes are natural catalysts, and although there would have been none in a pre-biotic world, there would have been catalytic minerals. Clay minerals form thin layers separated by a gap just wide enough to hold nucleotides, bringing them into close contact with each other and facilitating their union. Adding clay to nucleotide mixtures increases the length of the RNA chains that form spontaneously.[15]

The properties that give enzymes their catalytic abilities reside in the 3D shape of the protein, forming pockets or clefts in which other molecules are held and then joined or broken by molecular forces. RNA molecules called ribozymes also fold into 3D shapes and can catalyse reactions in the same way. In an RNA world, ribozymes could potentially replicate themselves or other kinds of RNA chain. Significantly for the RNA World hypothesis, the assembly of proteins by ribosomes depends on the catalytic ability of the RNA in the structure and not its protein component.

Once the primordial soup contained some spontaneously formed, random RNA chains one could expect that any that were able to copy themselves would multiply, but there is an obstacle to reaching this point. An individual RNA molecule cannot simultaneously be in the 3D shape required for catalysis and in the stretched out two-dimensional form required for copying. Thus, at least two identical molecules must cooperate; one has to be the template that will be copied and the other the catalyst that makes it happen. The chances of two such identical molecules meeting in a random RNA mixture and remaining together long enough to replicate are very low, but there is a solution.[16]

As we have seen before, cooperation is greatly facilitated if the cooperators can be corralled. In short, what is needed is a cell. This sounds like a catch-22, where to create the first cell one has first to have a cell, but we don't need anything nearly as complex as a cell membrane. All that is needed is a droplet of lipid. Lipids are oils and fats that, like so many biological molecules, are polymers, in this case made of fatty acids joined together. Most lipids on Earth today are biological in origin, but we now know that in a pre-biotic world lipid could have formed in the same reactions that produced nucleic acids, which is all very handy.

A primitive cell, or a lipid vesicle as we should really call it, would not only ensure that self-copying RNA molecules would be able to interact among themselves, but also that the copies they made would go on to replicate the same RNA sequences and not those of some other RNA that would not spread the self-copying gene. You may recognize that this is molecular kin selection.[17] At the time of writing, there has been no laboratory creation of an RNA molecule capable of unassisted self-replication as just described. It will be big news when it happens, as it surely will, because almost every other step in the RNA World scenario has now been synthesized. Most dramatically, lab experiments have shown how life can take off in an RNA world once the barrier to self-replication has been broken.

The lab system that demonstrated this take-off was first devised in the 1960s by Sol Spiegelman, a pioneer of molecular biology at the University of Illinois. Early in his career he had recognized that bacteria could provide ideal material for studying how genes work, a view not widely shared in the mid-1930s. His first paper was rejected by the editor of the American journal *Genetics* who commented that '[i]t is well known that bacteria have no nucleus and therefore can have no genetics'. The paper was finally published in the British *Journal of Genetics*, maybe because, Spiegelman thought, 'the English are so much more tolerant of deviants'.[18]

The first, simple version of Spiegelman's system used a stretch of RNA from a phage virus that infects bacteria. The RNA coded for an enzyme called a replicase that copied its own coded RNA sequence. This system could copy itself, so long as the RNA code could be translated into the enzyme, a protein. The phage would normally use the ribosomes of a host bacterium to do this, but it could be simulated in a test tube just by adding some purified replicase to the RNA and incubating it. What, Spiegelman wanted to know, 'will happen to the RNA molecules if the only demand made on them is the Biblical injunction, *multiply*, with the biological proviso that they do so as rapidly as possible?'

After 20 minutes of incubation, a sample of RNA was taken from the tube and transferred to a fresh tube with added replicase. After some more transfers, the incubation period was reduced to 15 minutes, then ten minutes, and finally just five minutes. With each reduction of incubation period there was selection for mutant, shorter RNA sequences whose replication could be completed in the time. By the 74th transfer, 83 per cent of the length of the RNA sequence had been lost. The RNA molecule had been reduced to just the short bit of sequence required for the replicase to recognize its template and no more.[19]

This simple replicase system demonstrated that RNA molecules could be made to evolve in a test tube by directed evolution. The newspapers ran stories about the creation of life in a test tube, but Spiegelman explained that he had done something different. 'When you create a living object the presumption is that the object didn't exist before. This I did not do. Working with simple chemical compounds, I take a primer of a living object and I generate many living objects from it.'

Fast-forward 50 years and the latest version of the replicase system has become a sophisticated RNA world in a tube. RNA replication still depends on the protein replicase, but importantly the enzyme can now evolve and not just its RNA template. The replicase is no longer added with each fresh transfer, but instead the mixture is fed with all the molecules needed for the RNA sequence to be translated into the enzyme in the tube itself. Now, each transfer inherits whatever replicase template has evolved from the previous generation. The RNA world is no longer supplied with an external source of functional replicase but contains only what evolves within the system.[20] What happens in such a genuinely self-replicating set-up?

Under directed evolution, shorter and shorter RNA sequences were selected as before but with an important difference. Lacking an external supply of replicase, those mutant RNA strands that became too short to produce functional replicase of their own used the enzyme produced by longer, more intact strands instead. This gave shorter strands an advantage because their smaller size enabled them to replicate more quickly. Parasitism by short strands on longer ones for replicase eventually crashed the whole system when unproductive parasites became too numerous. But there was a way to rescue the system.

The experiment was modified by adding oil droplets containing water to the tubes. These droplets, like primitive vesicles, provided compartments, each containing as before everything needed for RNA replication. Now, parasitic RNA could only spread within its own droplet, while intact RNA in droplets without parasites successfully and safely copied itself.

Compartmentalization protected cooperative RNA strands that supplied each other with replicase from extinction by parasitic RNA. When these experiments were run for hundreds of transfers, an unexpected thing happened.

Initially, two different RNA sequences evolved, each producing a replicase that worked only on itself. Then, one of the two sequences spawned a mutant that parasitized its progenitor. An evolutionary event occurring much faster but otherwise exactly like the Florideophyceae red algae which evolved daughter species that parasitize them. By 240 transfer generations, the host of the RNA parasite had spawned a second parasite and a third self-replicating lineage. There were now five main RNA lineages, linked together in a network by sharing or stealing replicase. All three replicase-producing lineages were now being parasitized, and replicase from one of the hosts was being used by everyone.[21]

Not only did parasites and cooperators evolve in this artificial RNA world, but one of the cooperators was indiscriminately altruistic, supplying vital replicase to all other lineages. Bearing in mind that we are talking about RNA molecules in lipid vesicles and not cells or organisms, the relationships that evolved are extraordinarily like those seen in complex life. The most singular result was revealed when the parasite that exploited all three self-replicating host lineages was removed. This caused the extinction of the altruist. The explanation for this is that the general parasite prevented the three host lineages from becoming so abundant that they competed strongly with each other. When that was allowed to happen by removing the general parasite, the altruist was the weakest of the three hosts, and it disappeared. It is hard to resist the conclusion that the presence of the altruist in this five-way community was maintained by interdependence among its members. That there can be an equivalent of the Mexican *Topos* in the RNA World is mind-blowing.

Almost as surprising as the lifelike evolution in these RNA experiments is that J. B. S. Haldane anticipated it 40 years before Spiegelman's experiments with phage. In a popular article he wrote for the *Rationalist Annual*, Haldane said in 1929:

> The first living or half-living things were probably large molecules synthesized under the influence of the Sun's radiation, and only capable of reproduction in the favourable medium in which they originated. Each presumably required a variety of highly specialised molecules before it could reproduce itself . . . This is the case today with most viruses, including the bacteriophage.[22]

John Maynard Smith, a former graduate student of Haldane, later joked when advising his own graduate students that you should always cite precedents, except in the case of Haldane who had illuminated so many problems that if they were all to be acknowledged, there would be nothing left to do.

In the same article on the origin of life, Haldane was the first to refer to the liquid medium in which life evolved as a 'soup'. This was scarcely his greatest contribution to science, but he coined a metaphor that captured the imagination and has been widely used ever since.[23] Haldane believed that the soup in which life emerged must have been a hot and dilute solution in the primitive ocean, but where exactly?

The location of the primordial soup kitchen has provoked lively debate. No one pretends that they know the exact spot, the way some people believe that they have located the final resting place of Noah's Ark. For one thing, more than three billion years of restless tectonic movement have probably erased any fossilized remains, though the oldest fossil environment so far found is more than 3.77 billion years old, so we shouldn't give up hope.[24] Research tells us quite precisely what we are looking for. Perhaps surprisingly, then, several different kinds of environment have been proposed as candidates, from the deep ocean to the seashore and hot springs.[25]

The first requirement is a source of energy. All cooking requires an input of energy to drive the reactions that turn ingredients into a product amenable to life, whether the dish is new or stew. Darwin imagined a warm little pond illuminated by the sun; Haldane agreed. The chemical reactions that produce nucleotides in the laboratory require UV light to purify them, also suggesting a surface environment for life's origin. Two other sources of energy, though not UV light, can be found in the deep ocean: energy produced by chemical reactions and geothermal heat.

The *Challenger* expedition found that the ocean hid deep canyons and tall mountains beneath its surface, but the significance of this submarine topography did not become apparent for another century. Then it was discovered that Earth's crust is composed of a series of tectonic plates that are thrust apart at the mid-oceanic ridges and subducted back into the mantle in deep ocean trenches. Fissures on the mid-ocean ridges vent hot water and dissolved minerals into the ocean, creating submarine oases that teem with weird life. Could the shattered dreams of the nineteenth-century scientists who hoped that they would find the origin of life in the deep now be fulfilled? That was certainly the thought when the vents were discovered.

In 1977 a deep-water submersible called *Alvin* was being used to explore the seabed near the Galapagos Islands in search of suspected hydrothermal vents. Tall chimneys were discovered, spewing forth clouds of hot water with the appearance of black smoke. These were dramatic enough, but more exciting still were the surreal animals picked out from the darkness by *Alvin*'s lights. Living in the cooler penumbra of the vents were giant tube-worms, some more than two metres long, with no mouth or anus; clams and mussels bigger than your head; and swarms of eyeless shrimp. What could feed the fauna of such a pitch-dark pressure cooker?

Far beyond the reach of sunlight, the base of the food web could not be photosynthesis, so it had to be its biochemical cousin: chemosynthesis. Photosynthesis liberates chemical energy through the reaction of oxygen and hydrogen, the latter obtained from splitting water molecules with the power of sunlight. In the hydrothermal vents live sulphur bacteria that are powered by the reaction of oxygen from seawater with hydrogen obtained from hydrogen sulphide issuing from the vents. Just as in photosynthesis, the chemical energy produced is used to make carbon compounds from carbon dioxide. Like plants do at the surface, the sulphur bacteria at hydrothermal vents feed everything else, either directly or indirectly. The molluscs filter bacteria from seawater, the shrimp feed on detritus, and the orificeless worms contain symbiotic bacteria that provide their needs.[26]

The ecosystem discovered around hydrothermal vents looked alien, but all its organisms have more familiar surface-living relatives, including sulphur bacteria found in the hot springs of Yellowstone. There was no *Urschleim*. The initial excitement around black smokers as a possible location for the origin of life cooled in favour a different kind of hydrothermal vent discovered at the turn of the millennium on the shoulders of mid-ocean ridges. These vents, some of them in shallower waters, are heated by chemical reactions in rocks as they are drenched in seawater that filters into fissures in the seabed. The water issuing from these vents is intensely alkaline and precipitates spongelike rock deposits that according to the alkaline vent theory (AVT) could have cradled the chemistry of first life.

At the time of writing, there is a tussle in progress between two camps espousing two different kinds of pre-biotic environment as the crucible where the building blocks of life—amino acids, nucleotides, and fatty acids—formed and became molecular cooperators. The AVT camp favour deep submarine vents, while the RNA camp favour shallow waters exposed to intense UV light. From the RNA World perspective, 'the idea that life originated at vents should, like the vents themselves, remain "In the deep bosom of the ocean buried" ', quoting Shakespeare's *Richard III*.[27]

The inventor of the AVT replied that the 'little pond people' are like the architect in the fictional realm visited by Gulliver in his travels who comes up with the clever idea to build houses from the roof down, ending with the foundations.[28] This is a smart-sounding debating point but a false analogy. The RNA World approach is starting from the roof in order to take the house to pieces so as to learn how it was built; it is not trying to build the house. But the true nub of the AVT argument is that when you get to the foundations in the ancient Earth of four billion years ago, there was only deep ocean and no land so there can have been no pond warmed by the sun. But we cannot be sure that there was no land at all—the merest scrap of a small island might have been enough. Eventually, this matter will be settled by evidence. For now, like Gulliver, we must move on. We note that, however it happened, life was born in chains, but these were biological polymers formed of cooperating units, not chains of bondage. How the polymer called DNA behaved is what interests us next.

14

Peas and justice

To the Augustinian friar Father Gregor Mendel, the peas he grew in the monastery garden at Brunn in Moravia were a source of amusement as well as science. His little joke was to announce to visitors, quite out of the blue and with a straight face, 'Now I am going to show you my children'. Doubtless he enjoyed the startled reaction of his guests to so frank an admission from a priest. If he had wanted to really shock them, he would have described his peas as '*hybriden*', which has the connotation 'bastards' in German, for hybrids they really were.

Mendel's garden was festooned with hundreds of pea plants, blooming white or purple, some with short stems and some tall. They climbed staves and the branches of trees and clung to strings stretched along the flower beds. Some flowers were swaddled in little paper or calico bags. These were the flowers that Mendel had used in his experimental crosses, using a camel hair brush to pollinate the flowers of one colour with the pollen of another and then protecting the flowers to prevent visiting insects from upsetting the cross. From such simple experiments blossomed the science of genetics, though not till 40 years after Mendel published the results and 16 years after his death.

Mendel's experiments showed that certain traits such as flower colour in his peas did not blend in the process of hybridization but were transmitted unchanged to future generations. A purple flower crossed with a white one did not produce pale purple offspring but pure purple ones. A further generation of crosses among these plants produced a mixture of offspring: three-quarters were purple, but a quarter were white. This was particulate, as distinct from blending, inheritance. The second-generation cross showed that white had not disappeared in the first cross, but was hidden, only to reappear.

The particles were eventually to be known as genes and were found to underlie the transmission of virtually all inherited traits, not just Mendelian ones. Traits such as human skin colour that appear to blend in the offspring of unalike parents are determined by many genes, each one with a small additive effect. Our modern understanding of how genes are transmitted is based on Mendel's discoveries, though the terminology, including the word 'gene' itself, is not his. Each gene comes in a variety of alternative versions, called alleles. Imagine a gene to be an ice cream cone that can carry one scoop of any available flavour. The flavours are alleles—you can have vanilla, chocolate, or strawberry—but only one flavour per cone and no mixture. But nature is nothing if not generous and she has given you two ice cream cones for each gene and allowed the second one to be a different flavour. In modern terms, what Mendel found was that when each parent contributes genes to its offspring, alleles (flavours) are represented in the next generation in proportion to their frequency in the parents.

For example, a white-flowered pea has two white alleles of the flower colour gene. In the first generation, all of Mendel's purple-flowered peas had two purple alleles. The frequency of the two alleles across all parents in the first cross between purple and white flowers was 50:50. All the offspring of this cross had one purple and one white allele—still half and half. But despite the presence of one white allele in all the offspring, none had white flowers because one purple allele is sufficient to make the purple pigment. Purple is said to be 'dominant' to white, which is in essence an allele unable to make pigment. At the next cross, the alleles were once again distributed according to their frequency in the parents. To have a white flower, a plant must receive two white alleles. The chance of this happening is $\frac{1}{2} \times \frac{1}{2}$ which is $\frac{1}{4}$. That is how Mendel ended up with a quarter of his peas having white flowers in the second-generation cross.

The last four paragraphs condense most of an entire lecture's worth of elementary genetics into 160 words, but for current purposes the important point can be summarized even more briefly in one short sentence: an allele is present among offspring in proportion to its frequency among parents. In two words, this is 'gene justice', or Mendel's First Law. The point being that the process of Mendelian inheritance is blind to the differences between alleles, as justice is said to be blind, and does not favour one allele over another. We shall see later how this kind of justice is a cornerstone of the evolution of cooperation among genes. But first we have a little more light spadework to do in Mendel's garden.

When Mendel examined the frequencies with which multiple traits were inherited, he found that these too behaved according to the statistical rule that allele frequencies in the parents are preserved in the offspring. So, when he counted the offspring of a cross involving the two traits flower colour (purple/white) and height of plant (tall/dwarf) the frequency of dwarf plants with white flowers was what would be predicted from randomly combining the individual allele frequencies in the parents. Here too, then, gene justice prevailed. But when, at the start of the twentieth century, Mendel's results were unearthed and scientists started replicating his experiments in other species and with many other traits, a complication appeared.

In 1905 at Cambridge University, William Bateson, Edith Saunders, and Reginald Punnett crossed sweet peas with the two traits in an experiment: flower colour (purple/red) and pollen shape (long/round). As expected, the allele frequencies were preserved among the offspring, but the combination of alleles was not what was expected. There were too many purple–long plants and too few red–long ones. These traits behaved as though their genes were in some way coupled, but an explanation as to why this was the case escaped Bateson. The answer came to Thomas Hunt Morgan and colleagues at Columbia University in New York.

Morgan conducted his genetics experiments with fruit flies, the little creatures that buzz around the compost bin and which are especially fond of ripe bananas. *Drosophila*, as the flies are technically known, can complete an entire generation in 12 days and labs breed them in their thousands in small bottles. The use of *Drosophila* was like rocket fuel to the emerging science of genetics, since it meant that large-scale experiments could be conducted very quickly. In his fly lab, Morgan discovered that just like in peas, some fly traits were coupled, but also that this coupling was sometimes broken. Using these results, Morgan made a leap that Bateson had steadfastly refused to take.

Morgan concluded that genes are carried on chromosomes—the finger-like structures visible in the nucleus of dividing cells and that were well known to microscopists (Figure 22). Bateson resisted the idea that chromosomes had anything to do with inheritance, but Morgan saw how the structure and behaviour of chromosomes could neatly explain the results of genetic experiments. Coupled genes must be physically located on the same chromosome.

During the cell division that produces gametes (sperm and eggs) chromosomes pair up, one from each parent. In this process, called meiosis, Mum's

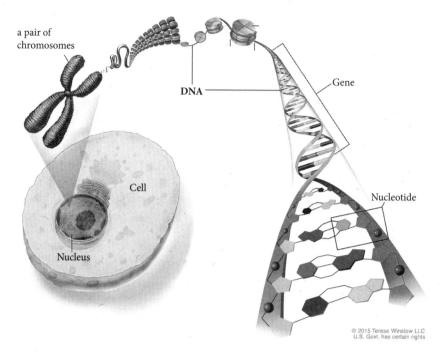

Figure 22 How DNA and genes are packaged into chromosomes

chromosome pairs with Dad's and they swap DNA. This process is called recombination. If the break where the swap occurs falls between two genes, they will be uncoupled during meiosis. The chance that this will occur depends on how far apart the genes are. If they are near each other, they will appear permanently linked, but if they are at opposite ends of the chromosome they will often be separated. Thomas Hunt Morgan received the Nobel Prize in 1933 for this insight, which was of monumental importance to the development of genetics.[1]

Why are genes joined together to make chromosomes? Such a question could scarcely have occurred to Morgan or other geneticists of his era, but it is an obvious one from a modern evolutionary perspective. Short stretches of RNA replicate faster, and as a consequence RNA molecules shrink under directed evolution, so it seems odd that even the handful of genes in a virus are joined to each other in a long molecule. Why should RNA or DNA molecules capable of self-replication link together in longer units? This is something of a trick question because there is a deception hidden in the phrase 'self-replication'. When it comes to self-replication, it takes at least two to tango: a template and a catalyst. The two could conceivably be doppelgangers, identical in sequence but performing different roles in the manner of ribozymes, though no such dual-function sequence has yet been discovered. Self-replicating molecules are like self-made millionaires: mythical. It takes cooperation with others to replicate, whether multiplying molecules or money.

The advantage of linkage between genes lies in their dependence of function. Two unlinked genes that depend on each other to perform their jobs in the cell or for successful replication risk being separated by cell division. Chromosomes solve this problem by providing a team bus, ensuring that no gene is left behind.[2] As cells evolved greater and greater complexity, genes formed larger and larger interdependent networks, and chromosomes would have grown in length, but this produced its own problem: errors in copying become inevitable in long messages.

The replicase that copies RNA in the phage experiments that we discussed in Chapter 13 can copy between 1 and 10,000 nucleotides without error, but this is only enough to code for a small enzyme.[3] The bacterium *E. coli* has a single circular chromosome composed of 4.6 million nucleotide pairs and most chromosomes typically carry thousands of genes, so they require error-correcting machinery. Evolution has obliged by equipping cells with several mechanisms that prevent and correct errors when DNA is copied.

To begin with, the chemical stability of DNA is superior to that of RNA and thus less prone to acquiring errors in the first place. This is presumably why DNA replaced RNA as the store for genetic information at an early date in the history of life. All cellular life uses it. The machinery for replicating DNA is similar in prokaryotes and eukaryotes, again suggesting an early date for its origin. The copying mechanism can be thought of as a single machine containing a dozen enzymes and other proteins that cooperate with each other. As a new DNA strand is built upon the template of an existing one by an enzyme called DNA polymerase, it proofreads the copy against the template and removes any mismatches, replacing them with the correct nucleotide.

Proofreading reduces the error rate from 1 in 100,000 before the check is made to 1 in 10 billion afterwards. Other enzymes follow on and carry out an additional check to repair any remaining mismatches. If this sounds like an anal level of error-checking, in one sense it is. A hereditary defect in mismatch repair produces a higher-than-normal accumulation of mutations in cells of the gut, resulting in a form of colon cancer.[4] J. B. S. Haldane died of colon cancer, which he typically scorned in a long piece of doggerel that included the lines 'My rectum is a serious loss to me / But I've a very neat colostomy'.

Despite all the error detection and correction that goes on when DNA is copied, some mutations do get through and establish themselves in the genome. It's just as well for evolution that they do because such mutation is the ultimate source of the genetic variation that produces novelty, and it provides the raw material on which natural selection acts. All the distinguishing features of any group of animals, plants, or microbes were once new.

Though mutations can be beneficial, they are much more often harmful for the simple reason that they occur randomly. What are the chances that a random change to any machine will improve its performance, after all? Just imagine: if random changes to a car could fix it, all repair manuals could be thrown away and car mechanics would become redundant. If, better still, random changes actually improved the car, automotive design in such a world would have more the appearance of *Mad Max* than Mercedes.

In the real world, where most mutations are harmful, there is a problem. To be sure, strongly deleterious mutations will be quickly removed by natural selection, but mildly harmful ones accumulate over the generations, eventually adding up to become a burden that reduces fitness. What can evolution do about this? In eukaryotes, which have the biggest chromosomes, the answer is sex and recombination.

Recombination during meiosis swaps pieces of chromosome with each other, and homologous pairs of chromosomes separate in different ways so that the composition of the genome is shuffled each generation. Over time, this patches defects caused by mutation, replacing them with intact genes from a different copy of the corresponding chromosome. In effect, recombination is a repair process.[5] This mechanism only works because the two copies of every chromosome in a cell are slightly different from each other, which can be relied upon because one came from Dad and the other from Mum.

Meiosis and sex go together like a horse and carriage. Meiosis is the horse that proverbially must come before the carriage. The function of meiosis is to reduce the two sets of chromosomes in a normal cell to the single set required to furnish gametes: eggs or sperm. When a gamete meets its match, egg and sperm fuse and the double set of chromosomes is restored. Under the microscope, meiosis is a complicated dance of the chromosomes. A rational choreographer commissioned to turn pairs of chromosomes into solo dancers would simply divide them in a single step $2 \to 1 + 1$. Meiosis does something much more complicated in which chromosomes multiply, pair, weave, and separate.

To begin, chromosomes double up, so that each has a conjoined twin sister. So instead of a single step of division, there is an initial increase: $2 \to 4$. Recombination then takes place and the chromosomes, still twinned, are dragged apart to opposite ends of the cell and separated by the formation of a new cell membrane. Finally, after a bit more dancing, the conjoined sisters separate and form gametes with a single chromosome set in each. Writing out the whole dance sequence in numbers of chromosome sets per cell, we have $2 \to 4 \to 2 \to 1$. Why has evolution made gamete production so unnecessarily complicated?

A likely answer is that the system protects gene justice by confounding alleles that try to cheat their way into more than their fair share of gametes. Cheating alleles called meiotic drivers occur widely, including in fruit flies, maize, and humans. One way they operate is by damaging competing chromosomes, to their own advantage, driving up their own frequency in the next generation. But thanks to meiosis, Mendel's First Law usually prevails. Indeed, meiosis has been called 'a trick to baffle the selfish genes'.[6] The trick is to thoroughly randomize which particular chromosome copies will end up together in the gametes so that cheats cannot rely on finding a susceptible target to damage. Randomization takes place each time the chromosomes are divided. Dividing just once, $2 \to 1$, would be the simple way, but by doubling first ($2 \to 4$) and then dividing twice ($4 \to 2 \to 1$), meiosis adds an additional level of uncertainty for cheats.

It is a general principle that uncertainty of outcome is the friend of justice and cooperation because it prevents cheats gaming the system.[7] This idea arises in the moral philosophy of John Rawls, who argued that a just society is one that is fair.[8] Decisions in a fair society should be made behind a 'veil of ignorance', he argued, as though we do not know how we would personally be affected by them. For example, to keep shares in a cake fair, the person who cuts it into slices should not be the one who decides which piece they get for themselves. Ignorance of who will get which piece encourages the person wielding the knife to divide the cake as exactly as they can into equal pieces.

Randomization during meiosis inserts a veil of ignorance into gamete production that keeps Mendelian inheritance fair. It has been suggested that politics could be made fairer by following this principle in the redrawing of

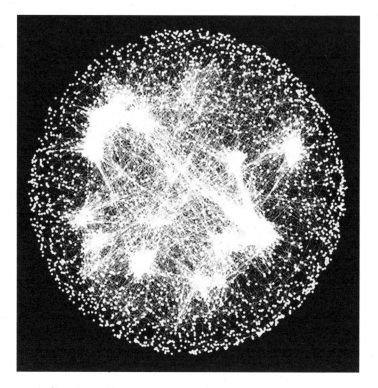

Figure 23 A diagrammatic representation of gene–gene interactions in yeast. Points represent individual genes and lines show which genes influence each other's function.

voting districts in the United States. Frequent randomizing of district boundaries would eliminate the anti-democratic practice called gerrymandering, in which the incumbent party redraws boundaries in its own favour.[9]

A political analogy has also been drawn between how decisions are reached in a parliamentary democracy and how genes make cooperative decisions. The biological question at issue is: how did a genome of thousands of individual genes, any of which could defect, come to cooperate in the evolution of the system of gene justice that is enacted in meiosis? Or, to put it more succinctly: how did cooperation triumph over cheating? There are two parts to the answer, reflecting the two essential ingredients of team building that have echoed throughout this book: a community of interest among team members and a mechanism to control cheats.

The community of interest among genes arises from the complexity of cellular processes. The level of cooperation required among molecular machines to replicate DNA represents an equivalent amount of cooperation among the genes that encode the proteins of which those machines are made. The same goes for all the other complex functions of the cell and the proteins that carry them out. The functional interdependence of processes in the cell makes the success and fitness of the genes in the cell interdependent too (Figure 23).

The genes in the genome all have a vested interest in the success of the individual that will transmit them to future generations. If the genome is imagined as a parliament of genes, they all vote for cooperation because it is in their individual interests to do so. And yet, we know that mutation can produce defectors who will try to cheat, as in cancer and meiotic drive. The reason that cheats are never successful for long is that they are outvoted by overwhelmingly superior numbers in the parliament of genes.[10] That doesn't stop the cheats trying though, as we are about to see.

15

Naked selfishness

Gene selfishness is not in itself a barrier to cooperation; in fact, it is the reverse. We have seen repeatedly how self-interested cooperation evolves because of the benefits it brings to team players. But cooperation does not just produce benefits, it also creates opportunities for cheats. Cheating reaches its most extreme in the genomic parasites that exploit the replication machinery of the cell. Viruses are a familiar example, and they are arguably nature's oldest professional cheats. There are RNA viruses in the sea with ancestors older than the LUCA, carrying genes encoding a replication enzyme that originates in the RNA world.[1] As is customary in long-established professions, viruses wear a uniform. Their protein coats provide identifiable targets for the human immune system. Other genomic cheats are harder to detect because they operate naked.

When the genome is described as a blueprint, an instruction manual, or a code book, the impression is given that it is an ordered store of relevant information, but it is in fact nothing like this. Most functional eukaryote genes—the ones that actually do something for the individual—do not exist as continuous sequences of meaningful genetic code but as fragments of sense DNA interspersed with fragments of nonsense. The meaningful fragments are known as exons, the meaningless ones are introns. Introns are genetic intruders, often consisting of long screeds of repetitive code that seem to serve no purpose whatsoever. Sometimes called junk DNA, these stretches of code are now recognized as genomic parasites called transposable elements (TEs) or transposons.

Over half our DNA is composed of TEs—parasitic DNA owns more of your genome than you do. Half of our TEs we share with mice and it has been sitting in our genome since our ancestors parted company with theirs 87 million years ago. The other half has accumulated since then. In plants, TEs can reach more than 95 per cent of the genome. It's a true wonder that

Figure 24 Barbara McClintock

genomes are decipherable and work at all. The variety of TEs is enormous and their nefarious strategies various—some are even parasitic within other TEs.[2] A eukaryote genome is part library and part zoo, with the unruly creatures of the menagerie constantly tearing into the pages, creating mutation and mayhem. The damage they cause is how TEs were first discovered.

Barbara McClintock (1902–1992) was the Mendel of maize (Figure 24). In a lifetime dedicated to genetic research and largely working alone, she used thousands of crosses to track the inheritance of traits from one generation to the next. The results enabled her to map genes to specific locations on the plant's ten chromosomes, which she was also able to visualize under the microscope. In 1948, she discovered that many of the genes she had been studying, such as those that determine the colour of maize grains, had recently become mutable, unexpectedly changing the appearance of grains. There appeared to be a gene in her self-pollinated maize stocks that moved around (transposed) on chromosome 9, causing a break where it reintegrated. Genes adjacent to the site of reintegration were inactivated, leading

McClintock to the insight that genes could switch each other off and on. She also discovered a second mobile gene that was required to activate the first.[3] Later, McClintock discovered that such genes could also move between chromosomes. We now know that they can move between species as well.

McClintock's discoveries were all made before DNA structure had been described by Watson and Crick or sequencing genes was possible. Later work showed that the two mobile genes, which we now know to be transposable elements, comprise one that encodes the enzyme transposase which allows the TE to jump location, and another that cannot produce transposase for itself. The latter gene depends upon the transposase produced by the former to enable it to move. We have seen this kind of dependence before in RNA sequences, commensal bacteria, and parasites. It also occurs routinely in almost all viruses. During an infection, cheating variants evolve that lack one or more genes essential to replication. These cheats depend for their spread on cooperative viruses that share enzymes and viral coat proteins as public goods.[4]

McClintock had predicted in 1950 that TEs would be found in all organisms. Time proved her right and brought her a belated Nobel Prize in 1983. TEs are highly diverse but have some features in common. Their sequences are mostly short and encode just a very limited number of proteins, all of which are involved in transposing the sequence. TEs are a source of mutation that can interfere with cell division and disable genes, causing disease. A TE inserting itself in the gene that codes for the blood-clotting protein Factor VIII results in haemophilia. Duchenne muscular dystrophy and predisposition to cancer are other conditions caused by TEs.[5]

Humans are of course not the only victims of TEs, nor is mutation the only damage TEs do. Pity the poor salamanders that appear to have been stupefied by the sheer exuberance of TEs that have caused their genomes to become bloated to 40 times the size of our own. The extra genome is not made of genes but of TEs, and this parasitic burden carries a cost. For reasons no one is sure of, big genomes make big cells. This applies in plants as well as animals, so whatever the cause of this correlation, it must be fundamental. The effect in salamanders is that their brain cells are huge. This does not make them smart but quite the opposite, because salamanders have limited cranial capacity so the price of filling it with large cells is that there are not many of them. Intelligence depends on lots of brain cells that are highly structured and deeply connected. Alas, salamanders are not playing with a full deck. But then nature does not demand they play poker, and they do all right if left alone by so-called more intelligent species.[6]

Eukaryote genomes are not alone in containing TEs; bacterial chromosomes also have them, but in bacteria they are usually deleted and not allowed by natural selection to accumulate. The exceptions are TEs that team up with genes that are of positive advantage to their bacterial host, for example in conferring antibiotic resistance. In contrast to bacteria, it is clear from the sheer numbers of TEs that accumulate in eukaryotes that deletion is not so ready an option. The reason boils down to the difference in generation time between bacteria (c. 20 minutes) and most eukaryotes (25 years in humans). Short generation time and huge population size in bacteria sharpen the blade of natural selection, but the reverse blunts its ability to remove TEs from eukaryote chromosomes.[7]

Though eukaryote cells have difficulty deleting TEs, the damage TEs can do has strongly selected for the evolution of countermeasures that prevent them from being transcribed. One of the most important ways cells do this is by attaching a small molecule called a methyl group to the DNA sequence. Methylated TE sequences are thereby mothballed and over generations they become inactivated by mutation, ending up as molecular fossils.

Methylation of DNA is used more generally to turn functional genes off and on and this kind of regulatory genetic switch is especially important during development. Methylation adds a control filter on top of the genes themselves, modifying their expression. Genetics plus methylation produces 'epigenetics'. It seems probable that DNA methylation evolved first as a means of controlling TEs, only later to become a means of managing a complex eukaryote genome through epigenetics.[8] If this is indeed how methylation first started, it is an example of a defence spin-off that knocks all others out of the park because development in multicellular organisms depends on epigenetics.

Epigenetics may be an indirect benefit of defence against genomic parasites, but TEs can occasionally have direct benefits too. After all, evolution would grind to a halt without mutations and TEs are a major source of these. Half the many mutations that *Drosophila* geneticists have discovered in their labs result from TEs. A textbook case of natural selection in the wild provides an example of a beneficial mutation.

The peppered moth (*Biston betularia*), found across the northern hemisphere, flies at night and by day hides on the undersides of the branches of trees (Figure 25). The wings and body of the moth are typically pale in

Figure 25 *Biston betularia,* typical (top) and carbonaria (bottom) forms

colour, peppered with flecks of black, but in the mid-nineteenth century British naturalists began to find previously unknown all-black, melanic forms. The frequency of melanism increased rapidly, reaching 90 per cent in the most polluted areas of England. In unpolluted areas, tree branches have lighter bark and are often covered in pale-coloured lichens, but lichens are absent from polluted areas and smoke turns trees black with soot. In one of the first demonstrations of natural selection operating in the wild, experiments showed that the spread of the melanic form of the peppered moth was the result of natural selection favouring moths that were better hidden from birds.[9] We now know that the mutation that turned peppered moths black was caused by the insertion of a large TE into a gene that affects colouration. The time that this occurred has even been dated to 1819, near the start of the coal-fired industrial revolution in Britain.[10]

The melanic mutation was a lucky strike from a TE spanner thrown into the genomic works. Rarely, TE spanners have become tools rather than projectiles, providing the genes for the evolution of some very important adaptations.[11] Like shoelaces that have caps on their ends to prevent them from fraying, chromosomes have caps on their ends called telomeres. With each replication, chromosomes shorten slightly at either end. Over a lifetime of cell division, this shortening could interfere with functional genes. Telomere sequences are highly repetitive, but this redundancy lends them resilience as the telomere gets shorter. How does a chromosome evolve a sacrificial, repetitive DNA sequence to protect its fraying ends? In some species, TEs provide such a thing ready-made. That will do nicely, thank you.[12]

TEs have also neatly solved another problem of eukaryote life. In the unceasing battle between eukaryotes and their hordes of bacterial and viral enemies, the parasites have the permanent advantage of short generation time, which lends them huge numbers and the capacity for rapid evolution. The difference so favours the parasites that it is unlikely that large multicellular organisms could have evolved at all without a solution to the problem. The solution is an immune system that can respond to pathogens on a timescale similar to their own, rather than at the glacial rate of evolution that constrains adaptation in the host. The system that evolved in vertebrates is called adaptive immunity.

Adaptive immunity works by first identifying the presence of an invader in the body and then tailoring and multiplying immune cells with specific receptors that seek out the pathogen and destroy it. The first step of identification is a tricky one because it requires specific molecular interaction with a potentially unlimited variety of invaders. The answer lies in combinatorics—that is, combining a limited number of variable molecules in every way possible to generate a variety of composites. For example, if we take a three-letter word, there are 26 possible letters of the English alphabet that we could place in the first position. Join each of those first letters to each of the 26 possibilities for the second letter and we have $26^2 = 676$, add a third letter and we get $26^3 = 17,576$. Of course, not all 17,576 words have a meaning in English, but if you are an immune cell that doesn't matter.

Adaptive immunity in vertebrates is a highly complex, exquisitely refined system with many variants. At its heart is a process of somatic recombination that mixes and matches the genes that make proteins called antibodies. The combinatorics of the system are so powerful that it produces an effectively unlimited variety of antibodies that between them are capable of attaching in a specific manner to almost any antigen (pathogen). Once a match has occurred between antibody and antigen, the cell type that makes that specific antibody multiplies clonally, producing a whole population of immune cells all churning out the antibody that is required. These tag the antigen, which is like drawing a target on the back of the invader, marking it for destruction.

If you want to rearrange just the specific genes that code for antibody proteins in immune cells, who you gonna call? Transposable elements. A pair of genes called RAG1 and RAG2 produce the enzymes that do the job. These genes have been permanent components of the vertebrate genome for 450 million years, originally belonging to a transposable element. In fact, they might be even older, since RAG1/2 have been found in the genome of the most primitive living chordate, a tiny animal called amphioxus, with which we share a common ancestor that lived more than 550 million years ago.[13]

Half a billion years is a very long time in which to allow a TE with the capacity to cause random mutations free range of the genome, however useful it may be. Natural selection has taken care of the problem, taming the transposable element so that it cannot cause mutation outside the limited part of the genome where its services are needed to build the basis of our adaptive immune system. In effect, the transposable element has been domesticated. Domestication is a familiar process, and in the context of symbiosis where we have met it before, it led to major transitions in evolution. Does the domestication of this TE constitute a major transition? I cannot see why not, even though it involved only a couple of genes joining team vertebrata. The adaptive immune system unquestionably depends on highly coordinated genetic teamwork.

Fruit flies, mosquitoes, and possibly other insects have a simple defence system against viruses that is augmented by an enzyme provided by TEs. Insect cells infected by a virus produce antiviral molecules, but these cannot on their own deal with viruses outside the infected cell. Active TEs produce an enzyme that copies the cellular antiviral into a form that is circulated around the entire body of the insect, providing life-saving protection. The enzyme in question is a public good shared between the TEs that produce it to enable their transposition. Insect cells that are host to TEs intercept the enzyme and use it for their own ends. Insects must tolerate the presence of active TEs and pay the price for this to receive protection from a worse threat. It's the original protection racket.[14]

Viruses—those jacketed villains of the genome—have on occasion been domesticated, particularly in a group of parasitic insects called parasitoid wasps. Parasitoid insects lay their eggs in or on the living bodies of other insects, where the parasitoid larvae hatch and consume the hapless host from the inside, usually reserving the vital organs as a final treat in order to keep the host alive as long as possible. In his autobiography, Darwin said that he thought this kind of suffering was a good argument against the existence of a beneficent creator. He didn't even know about viruses. Parasitoid wasps have teamed up with viruses, incorporating them into their own genomes. When laying an egg in a host, they inject the proteins made by viral genes into their victims where these assist the parasitoid by supressing the immune system or by altering the host's development and behaviour.[15]

Genes are promiscuous entities that form fluid alliances when this is favourable to their transmission. Over the four billion years that genes have existed, evolution has thrown up molecular tools for both getting genes into genomes and for getting rid of them. Genomic parasites such as transposons insert themselves into genomes using those tools. Other tools are used by cells to excise, destroy, or silence the intruders in defence of the community of interest determined by the parliament of their genes. Occasionally, when interests align, genomic parasites become members of the parliament.

All the conscious and unconscious efforts that people have made to shape species, from the domestication of the dog 30,000 years ago to the latest techniques of genetic engineering, use the molecular tools that evolution made and has been employing in nature for billions of years. This is not to say that genetic engineering should go unscrutinized. Rather, the point is that we need to look at the ethical and environmental consequences of engineering organisms and not focus on the supposed artificiality of the engineering process as activists have often done. For example, could we and should we engineer genomic parasites to force the extinction of species such as the mosquitoes that transmit malaria and dengue fever? Given the death and suffering caused by mosquito-borne diseases, the should-we question could be considered secondary to the could-we in this case. So, could the selfishness of genomic parasites endanger the very existence of a species? There is evidence that it might from an unexpected place.

Picture a sandstone bluff at the edge of a vast rocky plateau in the centre of the Sahara Desert in the south of Algeria. Years may go by without any rain and temperatures can fall below freezing at night and reach an unendurable 50°C in the day. Towering rocks blasted by the winds have formed fluted columns and fantastical, eroded shapes that are painted brick red by the setting sun. Scattered along the edge of the bluff, hidden in rocky defiles, and gathered in small groups in the beds of wadis is the only known population of the Sahara cypress *Cupressus dupreziana*, one of the most endangered trees on Earth. The entire population contains just 233 trees. A few are youngsters of 30, but the oldest are well over 2,000 years old and have survived many extreme climatic events.[16] But even these endurance records are not as remarkable as the unique sex life of the Sahara cypress. It is the only tree known to reproduce by androgenesis.

Many plants and some animals dispense with males, making seeds or eggs to which no male has contributed his genes. Only the Sahara cypress among plants and some marine bivalves among animals do the reverse, producing offspring that have a father and only a surrogate for a mother. Normally, meiosis produces pollen with a single set of chromosomes. Pollen of the Sahara cypress has a double set of chromosomes and has escaped the gene justice enforced by meiosis. Instead of fertilizing female cones, pollen in this species colonizes them, usurping the ovary and producing seeds containing an embryo that is a genetic clone of the father. This genetic takeover is known as gene drive. It drives a coach and horses through gene justice, to the advantage of genes transmitted exclusively via one sex—in this case male.

Abolishing males can be considered short-sighted, but abolishing females is surely suicidal. Genes lack foresight and are careless of the fate of species, so they can kill their carriers and thereby endanger a species for short-term gains in transmission. All that has saved Sahara cypress from this fate, at least so far, is that the trees are hermaphrodites. Every tree, even one with no genetic mother, produces female cones as well as male ones. Androgenetic bivalves are also hermaphrodites. How many, if any, non-hermaphrodite species may have been driven extinct by androgenesis in the course of evolutionary history we have no idea.

Androgenesis is just one of many mechanisms of gene drive, but it illustrates the feature common to all types, which is that selfish genes can and sometimes do spread through populations, even at the expense of the fitness of the individual. With that principle demonstrated by nature herself we can ask: how far can gene drive go if it is engineered? This is an area of active research. The first step is to engineer a gene sequence that can drive itself through a population by circumventing the normal mechanisms that enforce gene justice. That sequence must then damage the fitness of its carrier but do this in a way that does not select for resistance to the gene drive. All three of these steps are technologically very challenging, but there have been successes. Gene drives have been made that have wiped out large laboratory populations of malaria-transmitting mosquitoes.[17] How soon, if ever, such gene drives will be tested in nature and then used routinely to control disease and pests now depends on safeguards and a regulatory regime being agreed.[18]

16

Coda

The cornucopia of cooperation

Without cooperation there would be no life. Not just no life as we know it, but no life at all. The reason is simple. The essence of living is the ability to replicate and thereby to evolve. The simplest such entity is a self-replicating molecule. Its replication requires two copies to start: one to serve as a template and a second to catalyse the process. Although the two molecules are identical, they perform different functions: template and catalyst. So, even in this simplest of possible worlds, there must have been a division of labour between cooperating entities. As with the first self-reproducing molecules, so with all kinds of cooperators since the dawn of life: the first fundamental requirement for cooperation to evolve is that its benefits must outweigh its costs. A division of labour is a frequent route to such beneficial cooperation.

In a lonely and lifeless world, how do two copies of a rare molecule come into close association so that they can cooperate? Some kind of compartment is needed, such as a droplet of lipid forming a pre-biotic cell. In general, compartments favour cooperation because they bring cooperators together, enabling them to help each other, but compartments provide another advantage as well: cheat control. Cheats lose out because they are excluded or find themselves segregated in a compartment with their own kind: pickpockets at a pickpockets' convention. Cheat control is the second fundamental requirement for cooperation to evolve and, like the division of labour, it can be found present in some form or other from the dawn of life to the present (Table 2).

A large part of the cooperative history of life sketched in this book is about how the two fundamental requirements for cooperation, benefit, and cheat control have repeatedly played out over four billion years of

Table 2 Major transitions in evolution discussed in this book. Dates are million years ago (MYA). Sources can be found in the references given in the relevant chapters.

Major transition	MYA	Cooperators	Cheat control	Chapter
Origin of life and cells	4,000	RNA molecules	Compartmentation	13
Eukaryote chromosomes	>2,700	Genes	Two reduction divisions in meiosis	14
Mitochondria	2,700	Archaeon + bacterial ancestor of mitochondria	Uniparental (maternal) transmission of the mitochondrion	5, 10
Chloroplasts	1,250	Ancestor of green algae + cyanobacterium	Uniparental (maternal or paternal) transmission of the chloroplast	11
Complex multicellularity	1,000	Eukaryote cells	Passage through a unicell bottleneck every generation, cancer control genes, programmed cell death	12
Adaptive immunity	400	Vertebrate ancestor + RAG transposable elements	DNA methylation	15
Lichen symbiosis	250	Fungus + alga	Fungi control algal growth and in some lichens the alga is vertically transmitted	5
Eusociality in bees	113	Insect castes	Monogamy and kin selection	3
Fungus–farming ants	>50	Leafcutter ants + fungus	A single fungus clone that is vertically transmitted with ants	8

evolution (Table 2). This constancy over geological timescales is an instance of the uniformitarian principle that was divined in the rocks of Scotland by the eighteenth-century geologist James Hutton (1726–1797). Hutton established that everything visible in the record of the past can be explained by processes that can be seen operating in the present. In the age of Enlightenment, the uniformitarian principle unshackled reason from religion. There was no longer any need to believe in a lost age of miracles that fashioned creation and peopled the Earth in a never-to-be-repeated event. Without the flourishing of science stimulated by the Enlightenment, none of the evolutionary history of cooperation would be known.

Evolution is a gradual process—or, in the words of the eighteenth-century botanist Carl Linnaeus (1707–1778), *natura non facit saltus*, nature doesn't make jumps. Gradualism and uniformitarianism were essential to Darwin's argument in *The Origin of Species*, and yet there are major discontinuities in the history of life. The origin of life itself, the origin of eukaryotes, complex multicellularity, and all the other major transitions we have discussed are step changes during which nature made very big leaps indeed. The resolution of this apparent contradiction lies in how teams of cooperators are formed and then transformed at major transitions.

Teams are groups of cooperators, bound together by the benefits that accrue from the force of numbers and a division of labour. Once formed, teams may be transformed into a new kind of individual when the replication of team members becomes dependent on the replication of the team. This is how cells, eukaryotes, and multicellular organisms all arose, each constituting an MTE. MTEs forged by cooperation explain three fundamental features of biological organization: hierarchy, complexity, and emergence.

The hierarchy is the nested family of matryoshka, each hierarchical level represented by a doll. If we unpack the dolls representing a typical eukaryote and place them in a line, the very tiniest would represent the ribosome, a legacy of the RNA world. Then would be a bacterial cell, representing both the earliest of its kind and the ancestor of mitochondria. The next doll is the eukaryote cell that contains her. The fourth doll represents a multicellular individual and the fifth a eusocial colony such as a nest of ants or social bees.

Has the doll factory shut up shop, or is evolution still producing new major transitions? It has been claimed that the origin of our own species represents an MTE in its own right. I argued in Chapter 8 that using the criterion that major transitions produce a new kind of individual, the answer to this question must be 'no'. But a new MTE may be evolving among certain ants: the supercolony.[1] Normal ant colonies have territorial boundaries and ants attack rivals from other colonies. Supercolonies have no boundaries and bud new colonies containing individuals that move to and fro between colonies. Unlike the members of normal colonies, who are closely related to one another, the ants in supercolonies may be unrelated but still cooperate and move broods and resources between colonies as needed. They are still eusocial with sterile workers, but additionally there is communal breeding within supercolonies which contain many unrelated queens.[2]

Some supercolonial ants are spectacularly successful invaders, notably the Argentine ant that has spread to six continents and many oceanic islands. Despite a century of separation between populations of the invasive Argentine ant, when brought together in lab experiments ants treat each other as nestmates. This demonstrates that the worldwide invasion of the Argentine ant is one gigantic six-continent-sized supercolony.[3] The supercolony in the Mediterranean extends for 4,000 km across Spain, Portugal, France, and Italy (Figure 26). Is the supercolony a new kind of individual, reproducing as a unit and therefore qualifying as an MTE? It seems at least possible that it is, in which case the transformation of the supercolony seems to have been brought about through communal breeding built on top of kin selection and eusociality. If this sounds complex, it is!

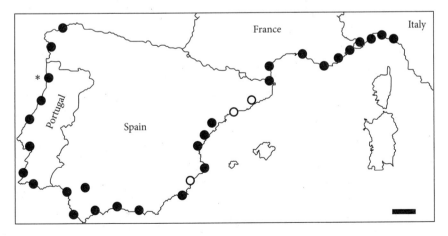

Figure 26 Map of a single Argentine ant supercolony (black dots) in southern Europe

Complexity increases during evolutionary history, but this has long been a controversial topic among biologists because of a hangover from the pre-Darwinian era.[4] Pre-Darwinian biology saw nature as arranged on a ladder, with lowly worms on the bottom rung of a Great Chain of Being and humans near the top, just below the angels. The French naturalist Jean-Baptiste Lamarck (1744–1829) was, like his contemporary Erasmus Darwin, an evolutionist, but he believed that the Great Chain of Being was the result of an inherent tendency in evolution towards progress. This made no sense at all to Charles Darwin, who wrote in his notebook, 'no higher or lower'.[5]

The Darwinian revolution replaced the ladder of ascent with a tree of descent, branching outwards from a single root. The appearance of progress is an illusion due to the fact that evolution has produced an accumulation of complexity. What we see in evolutionary history is not progress but an increase in the complexity of the individual arising from successive MTEs. A eukaryote cell that has teamed up with mitochondria is more complex than a prokaryote cell that has not. In terms of the metaphorical matryoshka, a nest of two dolls is more complex than a single doll. More complex organisms have evolved through more major transitions than simpler ones and contain more levels of organization.

In this context, 'level of organization' means the structures and functions found at a given point in the hierarchy. For example, at the cellular level we see bacteria communicating with each other through QS, sharing public goods, cross-feeding between species, and forming biofilms. Bacteria are social, but not to the same degree that multicellular organisms can be, with their cells specialized for communication and information processing. The neural cells dedicated to these tasks are specific to the multicellular level of organization and introduce a complexity of behaviour not seen among unicells. This brings us to the third feature of biological organization: emergence.

Emergence is a property of a system that is found when its parts interact with one another in a way that produces a result greater than the sum of the parts.[6] Recall the behaviour of slime moulds that feed as solitary amoebae in the soil in times of plenty, but which team up when the food runs out (Figure 19). The clarion call that assembles the team is a simple pulse of a signalling molecule that triggers any amoeba receiving it to migrate towards the source and to start signalling itself. The behaviour of solitary amoebae can be described in just three words—listen–signal–migrate—but what emerges when hundreds of thousands of them interact is a synchronized swarm with its own, more complex behaviour. Migrating cells fall into line with each other, marching in phalanx formation towards a point of convergence. At that point they form a multicellular slug that follows an extraordinary programme of development, producing a structure built from individual cells that is simultaneously quite unlike any of them.

Emergent phenomena, especially cooperative ones, are common in biology and can be found at all levels of organization. The difference between the 'tragedy of the commons' predicted by Garret Hardin, as described in Chapter 4, and the self-regulatory arrangements that Elinor Ostrom discovered operating among commoners sharing environmental public goods is a perfect example of emergence. Hardin thought that resources such as grazing land would be hit by heavier and heavier use until exhausted because each person would operate in selfish isolation. Ostrom showed that in actuality people often avoided this by instituting and policing environmental regulations that sustained the resource to everyone's benefit.

Operating on a longer timescale, the frequent evolution of cooperation between erstwhile parasites and hosts can be seen as emergent (Chapter 8). The linear impact of increasing parasite load on a host would be expected to result in the extinction of host, parasite, or both. Instead, cooperation, such as is found between microbes and insects in the bacteriome, emerges because endosymbionts and host share a community of interest in the survival of the host.

To return to the apparent paradox that selfish genes make social beings, this is ultimately resolved by the fact that the social rules of how we behave towards one another are emergent. When a major transition produces a new level of biological organization, new behaviours emerge. Emergent behaviours are not reducible to the properties of lower levels of organization. That is to say you cannot predict, for example, how social groups will behave purely from knowing how an individual's physiology works—even if that knowledge of physiology were impossibly complete. The reason is that social behaviour depends on the network of interactions among individuals, which is not purely a matter of individual physiology.

While social behaviour is emergent, it still has rules, and when it comes to a decision as to whether to cooperate or not, the rules are the familiar ones that operate at all levels. It does not usually pay to be selfish, even for genes. Cooperation yields numerous direct benefits to the individual, plus indirect ones through inclusive fitness. These two kinds of benefit are all that are required to explain the evolution of cooperation in nature, from people to bacteria. The siren call of group selection based on appeals to 'the good of the species' has in the past led thinking onto intellectual rocks that can now be avoided. There is no contradiction between the nature of the genes and the cooperation we choose to nurture in ourselves. That is not to say there is *no* conflict—there is plenty.

Cooperation survives in spite of conflict, not because it is absent. Conflict arises because cooperation always carries costs as well as benefits and cheats seek to profit without payment. This strategy may succeed in the short term, but it can be defeated or at least held in check by the fact that cheats need cooperators. Cooperators defeat cheats by surrounding themselves with other cooperators, ideally clones, and by policing to impose a cost on cheats. The ultimate in policing is when one member of the team controls the reproduction of the other, as happened when eukaryotes acquired mito-chondria, plants acquired chloroplasts, and humans domesticated dogs. Even in the case of domestication by humans, the process may not be intentional but rather the product of unconscious selection operating over many generations.

In the past, the matters covered in this book have, sometimes notoriously, been exploited as justifications and prescriptions for human behaviour. In the nineteenth and early twentieth centuries, social Darwinists saw justifica-tion for the status quo in evolution. Thankfully, Charles Darwin himself was not one of these. Peter Kropotkin drew opposite conclusions to those of the social Darwinists, but though celebrated in his day he is now unjustly for-gotten. My own political views make me highly sympathetic to arguments for cooperation, but I see no need to invoke biology to support them. Thanks to the major transitions in our evolutionary history, we are indi-vidually free to choose whether we cooperate or cheat. Just be warned, cheats rarely prosper or, to be exact, only prosper when rare.

Further reading

Cooperation is a huge subject but its coverage by works of popular science is patchy, with a superfluity of books about our own species and a dearth on other equally deserving topics. Here are titles that map approximately onto the four parts of the book, in which you will find some of the topics I have covered dealt with in more detail.

GROUPS

Bregman, R. (2020) *Humankind*. Bloomsbury, London.

Kohn, M. (2004) *A Reason for Everything: Natural Selection and the English Imagination*. Faber & Faber, London.

Nowak, M. A. & Highfield, R. (2012) *Super Cooperators: Beyond the Survival of Fittest—Why Cooperation, Not Competition, Is the Key to Life*. Canongate, Edinburgh.

Raihani, N. (2021) *The Social Instinct: How Cooperation Shaped the World*. Jonathan Cape, London.

Vince, G. (2019) *Transcendence: How Humans Evolved through Fire, Language, Beauty, and Time*. Allen Lane, London.

INDIVIDUALS

Alcock, J. (2009) *The Triumph of Sociobiology*. Oxford University Press, Oxford.

Archibald, J. (2014) *One Plus, One Equals, One: Symbiosis and the Evolution of Complex Life*. Oxford University Press, Oxford.

Maynard Smith, J. & Szathmáry, E. (2000) *The Origins of Life: From the Birth of Life to the Origin of Language*. Oxford University Press, Oxford.

CELLS

Aktipis, A. (2021) *Cheating Cell: How Evolution Helps Us Understand and Treat Cancer*. Princeton University Press, Princeton, NJ.

Brasier, M. (2012) *Secret Chambers: The Inside Story of Cells and Complex Life*. Oxford University Press, Oxford.

Lane, N. (2016) *The Vital Question: Why Is Life the Way It Is?* Profile, London.

Mukherjee, S. (2022) *The Song of the Cell: An Exploration of Medicine and the New Human.* Random House, London.

GENES

Mukherjee, S. (2017) *The Gene: An Intimate History.* Penguin Random House, London.

Ramakrishnan, V. (2018) *Gene Machine: The Race to Decipher the Secrets of the Ribosome.* One World Publications, London.

Yanai, I. & Martin, L. (2016) *The Society of Genes.* Harvard University Press, Cambridge, MA.

Acknowledgements

Many have cooperated in the production of this book. I thank the colleagues and friends who read the manuscript at various stages of preparation, especially Rob Freckleton, Bridget Laue, David Leach, Caroline Pond, Steven Rose, Menno Schilthuizen, and Dave Wilkinson, who read all or most of the book. I am grateful to colleagues David Finnegan, Sandy Hetherington, Thorun Helgason, and Richard Milne, who gave me the benefit of their expertise in particular chapters. It is a cliché to say that any remaining mistakes are mine, but it is true nonetheless. I also take full responsibility for the jokes.

I am fortunate to belong to an institute in Edinburgh that is replete with experts in most branches of evolutionary biology, and I have benefitted from conversations with all of them, especially Matt Bell, Nick Colegrave, and Per Smiseth. It is a famously collegial bunch of people, both in the Darwin Dance Hall (our coffee room) and on the picket line. I wish that the struggle of UK academics for fair treatment and better pay was temporary, but it has been a constant throughout my career, which is why I have dedicated the book as I have.

At Oxford University Press I am indebted to my editor Latha Menon and to Tara Werger who managed book production. My agent, Peter Tallack, helped shape the book into a saleable commodity—thank you, Peter. Finally, I must acknowledge the contribution of my kin. I have learned a great deal from my sons Dan and Ben, who have embodied the principles and practice of cooperation in their joint enterprise. My wife, Rissa de la Paz, has read this book with the same critical eye she has applied to all my previous work, but I cannot measure what her support means to me in mere words.

Endnotes

CHAPTER I

1. https://en.wikipedia.org/wiki/1985_Mexico_City_earthquake [Accessed 4.1.21].
2. https://www.topos.mx/formacion/aspirantes [Accessed 4.1.21].
3. https://www.bbc.co.uk/news/world-latin-america-53332756 [Accessed 4.1.21].
4. https://edition.cnn.com/2020/06/05/americas/mexico-femicide-coronavirus-lopez-obrador-intl/index.html [Accessed 4.1.21].
5. https://www.macrotrends.net/countries/ranking/murder-homicide-rate [Accessed 5.6.23].
6. Bregman, R. (2020) *Humankind*. Bloomsbury, London.
7. Bowles, S. & Gintis, H. (2013) *A Cooperative Species: Human Reciprocity and Its Evolution*. Princeton University Press, Princeton; Nowak, M. A. & Highfield, R. (2012) *Super Cooperators: Beyond the Survival of Fittest—Why Cooperation, Not Competition, Is the Key to Life*. Canongate, Edinburgh; Henrich, J. & Muthukrishna, M. (2021) The Origins and Psychology of Human Cooperation. *Annual Review of Psychology*, 72, 207–240.
8. Dawkins, R. (2006) *The Selfish Gene*. 30th Anniversary edition. Oxford University Press, Oxford.
9. A note on dates. The dates given in this history range from precisely known facts to ancient evolutionary events that can only be estimated to within hundreds of millions of years. These estimates are subject to constant revision. But an author has one date he cannot ignore: publication day. Therefore, all evolutionary dates given are the most plausible at the time of writing and some will inevitably prove to be wrong. For our purposes, the order of events is more important than precise dates, which is why I have also omitted date ranges unless this is relevant to the discussion. Readers interested in more precision will usually find this in the references cited.
10. https://www.britannica.com/event/World-War-I/Killed-wounded-and-missing [Accessed 26/4/21].
11. https://www.iwm.org.uk/history/voices-of-the-first-world-war-the-christmas-truce [Accessed 26/4/21].
12. Ashworth, T. (1980) *Trench Warfare 1914–1918: The Live and Let Live System*. Macmillan, London.
13. Ashworth, T. (1980) *Trench Warfare 1914–1918: The Live and Let Live System*. Macmillan, London.

14. Ashworth, T. (1980) *Trench Warfare 1914–1918: The Live and Let Live System.* Macmillan, London.

15. Ashworth, T. (1980) *Trench Warfare 1914–1918: The Live and Let Live system.* Macmillan, London.

16. Axelrod, R. (1990) *The Evolution of Cooperation.* Penguin Books, London.

CHAPTER 2

1. Woodcock, G. & Avakumović, I. (1950) *The Anarchist Prince: A Biographical Study of Peter Kropotkin.* T. V. Boardman, London; New York.

2. Woodcock, G. & Avakumović, I. (1950) *The Anarchist Prince: A Biographical Study of Peter Kropotkin.* T. V. Boardman, London; New York.

3. Kropotkin, P. A. (1902) *Mutual Aid, a Factor of Evolution.* Garland Pub., New York.

4. Todes, D. P. (1989) *Darwin without Malthus.* Oxford University Press, Oxford.

5. Malthus, T. R. (1970) *An Essay on the Principle of Population; and, A Summary View of the Principle of Population.* Penguin, Harmondsworth.

6. Huxley, T. H. (1888) The struggle for existence in human society. *The Nineteenth Century.* February 1888 issue. As quoted in: Dugatkin, L. A. (2013) Strange bedfellows: A Russian prince, a Scottish economist, and the role of empathy in early theories for the evolution of cooperation. *Journal of Experimental Zoology,* 320, 407–411.

7. Kropotkin, P. A. (1902) *Mutual Aid, a Factor of Evolution.* Garland Pub., New York.

8. Todes, D. P. (1989) *Darwin without Malthus.* Oxford University Press, Oxford, p. 137.

9. Shukla, S. P. et al. (2018) Microbiome-assisted carrion preservation aids larval development in a burying beetle. *Proceedings of the National Academy of Sciences,* 115, 11274–11279.

10. Scott, M. P. (1998) The ecology and behavior of burying beetles. *Annual Review of Entomology,* 43, 595–618.

11. Chen, B. F., Liu, M., Rubenstein, D. R., Sun, S. J., Liu, J. N., Lin, Y. H., & Shen, S. F. (2020) A chemically triggered transition from conflict to cooperation in burying beetles. *Ecology Letters,* 23, 467–475.

CHAPTER 3

1. Leeson, P. T. (2007) An-arrgh-chy: The law and economics of pirate organization. *Journal of Political Economy,* 115, 1049–1094.

2. Leeson, P. T. (2007) An-arrgh-chy: The law and economics of pirate organization. *Journal of Political Economy,* 115, 1049–1094.

3. Munro, C., Vue, Z., Behringer, R. R., & Dunn, C. W. (2019) Morphology and development of the Portuguese man of war, Physalia physalis. *Scientific Reports,* 9, https://doi.org/10.1038/s41598-019-51842-1.

4. Hamilton, W. D. (1964) Genetical evolution of social behaviour I. *Journal of Theoretical Biology,* 7, 1–16; Hamilton, W. D. (1964) Genetical evolution of social behaviour II. *Journal of Theoretical Biology,* 7, 17–52.

5. There is one case known in nature, the Tibetan ground tit, where the two sides of Hamilton's formula applied to helpers at the nest are in fact equal. In such a situation, it is predicted that helping behaviour will be polymorphic, with some birds helping and others not doing so, depending on a heritable disposition towards one behaviour or the other. This prediction is confirmed in Wang, C. C. & Lu, X. (2018a) Hamilton's inclusive fitness maintains heritable altruism polymorphism through rb = c. *Proceedings of the National Academy of Sciences*, 115, 1860–1864 and Wang, C. C. & Lu, X. (2018b) Inclusive fitness does maintain a heritable altruism polymorphism in Tibetan ground tits REPLY. *Proceedings of the National Academy of Sciences*, 115, E11210–E11211.

6. Maynard Smith, J. (1964) Group selection and kin selection. *Nature (London)*, 201, 1145–1147.

7. Brueland, H. (1995) Chapter 18: Highest Lifetime Fecundity. The University of Florida Book of Insect Records, Department of Entomology and Nematology, UF/IFAS. https://entnemdept.ufl.edu/walker/ufbir/chapters/chapter_18.shtml [Accessed 11.7.22].

8. Hughes, W. O. H. et al. (2008) Ancestral monogamy shows kin selection is key to the evolution of eusociality. *Science*, 320, 1213–1216.

9. Faulkes, C. G. & Bennett, N. C. (2021) Social evolution in African mole-rats: a comparative overview. In *Extraordinary Biology of the Naked Mole-Rat*. Springer Nature AG, Cham, Switzerland (eds R. Buffenstein, T. J. Park & M. M. Holmes), pp. 1–33.

10. Clutton-Brock, T. (2021) Social evolution in mammals. *Science*, 373, eabc9699. https://doi.org/10.1126/science.abc9699. Burda, H. et al. (2000) Are naked and common mole-rats eusocial and if so, why? *Behavioral Ecology and Sociobiology*, 47, 293–303.

11. Buffenstein, R. (2021) Colony-specific dialects of naked mole-rats. *Science*, 371, 461–462.

12. Cockburn, A. (2006) Prevalence of different modes of parental care in birds. *Proceedings of the Royal Society B: Biological Sciences*, 273, 1375–1383.

13. Downing, P. A., Griffin, A. S., & Cornwallis, C. K. (2020) The benefits of help in cooperative birds: nonexistent or difficult to detect? *American Naturalist*, 195, 1085–1091.

14. Green, J. P., Freckleton, R. P., & Hatchwell, B. J. (2016) Variation in helper effort among cooperatively breeding bird species is consistent with Hamilton's rule. *Nature Communications*, 7. https://doi.org/10.1038/ncomms12663.

15. Hatchwell, B. J., Gullett, P. R., & Adams, M. J. (2014) Helping in cooperatively breeding long-tailed tits: a test of Hamilton's rule. *Philosophical Transactions of the Royal Society B: Biological Sciences*, 369. https://doi.org/10.1098/rstb.2013.0565.

16. I have rounded the values of r, b, and c presented here and simplified the calculations for the sake of readability, but the conclusion is the same as reached in the original paper of Hatchwell, B. J., Gullett, P. R., & Adams, M. J. (2014) Helping in cooperatively breeding long-tailed tits: a test of Hamilton's rule. *Philosophical Transactions of the Royal Society B: Biological Sciences*, 369. https://doi.org/10.1098/rstb.2013.0565.

17. Murton, R. K. (1971) *Man and Birds.* Collins, London.
18. Baglione, V. & Canestrari, D. (2016) Carrion crows: Family living and helping in a flexible social system. In *Cooperative Breeding in Vertebrates: Studies of Ecology, Evolution, and Behavior* (eds J. L. Dickinson & W. D. Koenig). Cambridge University Press, Cambridge, pp. 97–114.
19. Griffin, A. S. & West, S. A. (2003) Kin discrimination and the benefit of helping in cooperatively breeding vertebrates. *Science*, 302, 634–636.
20. Garcia-Ruiz, I., Quinones, A., & Taborsky, M. (2022) The evolution of cooperative breeding by direct and indirect fitness effects. *Science Advances*, 8. https://doi.org/10.1126/sciadv.abl7853.
21. Cornwallis, C. K. et al. (2017) Cooperation facilitates the colonization of harsh environments. *Nature Ecology & Evolution*, 1, 57.
22. Cornwallis, C. K. et al. (2010) Promiscuity and the evolutionary transition to complex societies. *Nature*, 466, 969–972.
23. Lukas, D. & Clutton-Brock, T. (2018) Social complexity and kinship in animal societies. *Ecology Letters*, 21, 1129–1134; Federico, V. et al. (2020) Evolutionary pathways to communal and cooperative breeding in carnivores. *American Naturalist*, 195, 1037–1055.
24. Bourke, A. F. G. (2021) The role and rule of relatedness in altruism. *Nature*, 590, 392–394.

CHAPTER 4

1. Clark, R. W. (1984) *J. B. S., the Life and Work of J. B. S. Haldane.* Oxford University Press, Oxford.
2. Haldane, J. B. S., Sprunt, A. D., & Haldane, N. M. (1915) Reduplication in mice (preliminary communication). *Journal of Genetics*, 5, 133–135.
3. Raihani, N. J. & Bshary, R. (2015) Why humans might help strangers. *Frontiers in Behavioral Neuroscience*, 9, 1–11.
4. Lents, N. H. (2020) *Human Errors: A Panorama of Our Glitches, from Pointless Bones to Broken Genes.* Weidenfeld & Nicolson, London.
5. Zuk, M. (2013) *Paleofantasy: What Evolution Really Tells Us about Sex, Diet, and How We Live.* W. W. Norton, New York.
6. Ardrey, R. (1967) *The Territorial Imperative.* Atheneum, London. For an alternative view from the same period, see Montagu, M. F. A. (1968) *Man and Aggression.* Oxford University Press, New York.
7. Burkart, J. M. et al. (2014) The evolutionary origin of human hyper-cooperation. *Nature Communications*, 5. https://doi.org/10.1038/ncomms5747.
8. Ostwald, M. M., Haney, B. R., & Fewell, J. H. (2022) Ecological drivers of non-kin cooperation in the hymenoptera. *Frontiers in Ecology and Evolution*, 10. https://doi.org/10.3389/fevo.2022.768392.
9. Ostwald, M. M. et al. (2021) Cooperation among unrelated ant queens provides persistent growth and survival benefits during colony ontogeny. *Scientific Reports*, 11. https://doi.org/10.1038/s41598-021-87797-5.

10. Riehl, C. (2010) A simple rule reduces costs of extragroup parasitism in a communally breeding bird. *Current Biology*, 20, 1830–1833.

11. Raihani, N. (2021) *The Social Instinct: How Cooperation Shaped the World*. Jonathan Cape, London.

12. Viana, D. S. et al. (2010) Cognitive and motivational requirements for the emergence of cooperation in a rat social game. *Plos One*, 5. https://doi.org/ 10.1371/ journal.pone.0008483.

13. Schweinfurth, M. K. & Taborsky, M. (2018) Norway rats (*Rattus norvegicus*) communicate need, which elicits donation of food. *Journal of Comparative Psychology*, 132, 119–129.

14. Sparks, A., Burleigh, T., & Barclay, P. (2016) We can see inside: accurate prediction of prisoner's dilemma decisions in announced games following a face-to-face interaction. *Evolution and Human Behavior*, 37, 210–216.

15. Noe, R. (2006) Cooperation experiments: coordination through communication versus acting apart together. *Animal Behaviour*, 71, 1–18; Raihani, N. J. & Bshary, R. (2011) Resolving the iterated prisoner's dilemma: theory and reality. *Journal of Evolutionary Biology*, 24, 1628–1639.

16. Tomasello, M. et al. (2012) Two key steps in the evolution of human cooperation: the interdependence hypothesis. *Current Anthropology*, 53, 673–692.

17. Hrdy, S. B. (2011) *Mothers and others: the evolutionary origins of mutual understanding*. Belknap Press of Harvard University Press, Cambridge, MA.

18. Hill, K. R. et al. (2011) Co-residence patterns in hunter-gatherer societies show unique human social structure. *Science*, 331, 1286–1289.

19. Boyd, R. & Richerson, P. J. (2021) Large-scale cooperation in small-scale foraging societies. *Evolutionary Anthropology*. https://doi.org/ 10.1002/evan.21944.

20. Roe, F. G. (1972) *The North American Buffalo: A Critical Study of the Species in Its Wild State*. David & Charles, Newton Abbot.

21. John Fire Lame Deer on the importance of buffalo to Native Americans: https://en.wikipedia.org/wiki/Bison [Accessed 31/07/22].

22. Taylor, W. T. T. et al. (2023) Early dispersal of domestic horses into the Great Plains and northern Rockies. *Science*, 379, 1316–1323.

23. Brink, J. W. (2014) *Imagining Head-Smashed-In: Aboriginal Buffalo Hunting on the Northern Plains*. AU Press, Edmonton.

24. Boyd, R. & Richerson, P. J. (2021) Large-scale cooperation in small-scale foraging societies. *Evolutionary Anthropology*. https://doi.org/ 10.1002/evan.21944.

25. Boyd, R. & Richerson, P. J. (2021) Large-scale cooperation in small-scale foraging societies. *Evolutionary Anthropology*. https://doi.org/ 10.1002/ evan.21944.

26. Tomasello, M. et al. (2012) Two key steps in the evolution of human cooperation: the interdependence hypothesis. *Current Anthropology*, 53, 673–692.

27. Domínguez-Rodrigo, M. et al. (2014) On meat eating and human evolution: a taphonomic analysis of BK4b (Upper Bed II, Olduvai Gorge, Tanzania), and its bearing on hominin megafaunal consumption. *Quaternary International*, 322–323, 129–152; Surovell, T., Waguespack, N., & Brantingham, P. J. (2005)

Global archaeological evidence for proboscidean overkill. *Proceedings of the National Academy of Sciences*, 102, 6231–6236.

28. Apicella, C. L. et al. (2012) Social networks and cooperation in hunter-gatherers. *Nature*, 481, 497–501; Apicella, C. L. & Silk, J. B. (2019) The evolution of human cooperation. *Current Biology*, 29, R447–R450.

29. Rand, D. G., Arbesman, S., & Christakis, N. A. (2011) Dynamic social networks promote cooperation in experiments with humans. *Proceedings of the National Academy of Sciences*, 108, 19193–19198.

30. Clutton-Brock, T. H. & Parker, G. A. (1995) Punishment in animal societies. *Nature*, 373, 209–216. Silk, J. B. (2007) Chimps don't just get mad, they get even. *Proceedings of the National Academy of Sciences*, 104, 13537–13538; Fehr, E. & Gachter, S. (2002) Altruistic punishment in humans. *Nature*, 415, 137–140; Vaish, A. et al. (2016) Preschoolers value those who sanction non-cooperators. *Cognition*, 153, 43–51; Ågren, J. A., Davies, N. G., & Foster, K. R. (2019) Enforcement is central to the evolution of cooperation. *Nature Ecology & Evolution*, 3, 1018–1029.

31. Darwin, *The Descent of Man*, 2nd ed. p. 201.

32. *Othello*, Scene 3, Act 3.

33. Brears, R. C. (2022) *Debt-for-Nature Swaps Financing Nature-Based Solutions—Financing Nature-Based Solutions: Exploring Public, Private, and Blended Finance Models and Case Studies*. Springer International Publishing, Cham, pp. 51–73.

34. Ostrom, E. (2015) *Governing the Commons: The Evolution of Institutions for Collective Action*. Cambridge University Press, Cambridge.

CHAPTER 5

1. Sapp, J. (1994) *Evolution by Association: A History of Symbiosis*. Oxford University Press, New York.

2. Reproduced in King-Hele, D. (1968) *Essential Writings of Erasmus Darwin*. McGibbon & Kee, London, pp. 90–91.

3. Brown, J. (1995) *Charles Darwin: Voyaging*. Pimlico, London.

4. Sapp, J. (1994) *Evolution by Association: A History of Symbiosis*. Oxford University Press, New York, ch. 1.

5. https://en.wikipedia.org/wiki/Timeline_of_abolition_of_slavery_and_serfdom [Accessed 29.03.2022].

6. Oulhen, N., Schulz, B. J., & Carrier, T. J. (2016) English translation of Heinrich Anton de Bary's 1878 speech, 'Die Erscheinung der Symbiose' ('De la symbiose'). *Symbiosis*, 69, 131–139.

7. Unfortunately, 'symbiosis' is not infrequently used as a synonym for mutualism. See Wilkinson, D. M. (2001). At cross purposes. *Nature*, 412, 485.

8. Sapp, J. (1994) *Evolution by association: A history of symbiosis*. Oxford University Press, New York.

9. Sapp, J. (1994) *Evolution by association: A history of symbiosis*. Oxford University Press, New York.

10. Buchner, P. (1930) *Tier und Pflanze in Symbiose*. Gebrüder Borntraeger, Berlin.

11. Sagan, L. (1967) On the origin of mitosing cells. *Journal of Theoretical Biology*, 14, 255–274.

12. Watson, J. D. & Crick, F. H. (1953) Genetical implications of the structure of deoxyribonucleic acid. *Nature*, 171, 964–967.

13. Watson, J. D. & Crick, F. H. (1953) Genetical implications of the structure of deoxyribonucleic acid. *Nature*, 171, 964–967.

14. Stocking, C. R. & Gifford, E. M. (1959) Incorporation of thymidine into chloroplasts of Spirogyra. *Biochemical and Biophysical Research Communications*, 1, 159–164.

15. Margulis, L. (2005) Hans Ris (1914–2004): genophore, chromosomes and the bacterial origin of chloroplasts. *International Microbiology*, 8, 145–148.

16. Sagan, L. (1967) On the origin of mitosing cells. *Journal of Theoretical Biology*, 14, 255–274.

17. Gibson, T. M. et al. (2018) Precise age of *Bangiomorpha pubescens* dates the origin of eukaryotic photosynthesis. *Geology*, 46, 135–138.

18. Field, C. B. et al. (1998) Primary production of the biosphere: integrating terrestrial and oceanic components. *Science*, 281, 237–240.

19. Maynard Smith, J. & Szathmáry, E. (1995) *The Major Transitions in Evolution*. W. H. Freeman, Oxford.

20. Herron, M. D. (2021) What are the major transitions? *Biology & Philosophy*, 36. https://doi.org/10.1007/s10539-020-09773-z.

21. West, S. A., Fisher, R. M., Gardner, A., & Kiers, E. T. (2015) Major evolutionary transitions in individuality. *Proceedings of the National Academy of Sciences*, 112, 10112–10119.

CHAPTER 6

1. Lucking, R. et al. (2014) A single macrolichen constitutes hundreds of unrecognized species. *Proceedings of the National Academy of Sciences*, 111, 11091–11096.

2. Lucking, R. et al. (2009) Do lichens domesticate photobionts like farmers domesticate crops? Evidence from a previously unrecognized lineage of filamentous cyanobacteria. *American Journal of Botany*, 96, 1409–1418.

3. Tripp, E. A. & Lendemer, J. C. (2018) Twenty-seven modes of reproduction in the obligate lichen symbiosis. *Brittonia*, 70, 1–14.

4. Visick, K. L., Stabb, E. V., & Ruby, E. G. (2021) A lasting symbiosis: how *Vibrio fischeri* finds a squid partner and persists within its natural host. *Nature Reviews Microbiology*, 19, 654–665. Nyholm, S. V. & McFall-Ngai, M. J. (2021) A lasting symbiosis: how the Hawaiian bobtail squid finds and keeps its bioluminescent bacterial partner. *Nature Reviews Microbiology*, 19, 666–679.

5. Nealson, K. H., Platt, T., & Hastings, J. W. (1970) Cellular control of the synthesis and activity of the bacterial luminescent system. *Journal of Bacteriology*, 104, 313–322.

6. Dunlap, P. V. et al. (2007) Phylogenetic analysis of host–symbiont specificity and codivergence in bioluminescent symbioses. *Cladistics*, 23, 507–532.

7. Chomicki, G. et al. (2020) Compartmentalization drives the evolution of symbiotic cooperation. *Philosophical Transactions of the Royal Society B: Biological Sciences*, 375, 20190602. https://doi.org/10.1098/rstb.2019.0602.

8. Campbell, A. K. (2008) Jean-Marie Bassot (1933–2007): a life of unquenched curiosity—Obituary. *Luminescence*, 23, 187–190.

9. Douglas, A. E. (2015) Multiorganismal insects: diversity and function of resident microorganisms. *Annual Review of Entomology*, 60, 17–34.

10. Aanen, D. K. & Eggleton, P. (2017) Symbiogenesis: beyond the endosymbiosis theory? *Journal of Theoretical Biology*, 434, 99–103.

11. Douglas, A. E. (2015) Multiorganismal insects: diversity and function of resident microorganisms. *Annual Review of Entomology*, 60, 17–34.

12. Hammer, T. J., Sanders, J. G., & Fierer, N. (2019) Not all animals need a microbiome. *Fems Microbiology Letters*, 366. https://doi.org/10.1093/femsle/fnz117.

13. Maire, J. et al. (2018) An IMD-like pathway mediates both endosymbiont control and host immunity in the cereal weevil Sitophilus spp. *Microbiome*, 6, 6.

14. Garcia, J. R. & Gerardo, N. M. (2014) The symbiont side of symbiosis: do microbes really benefit? *Front Microbiol*, 5, 510; Keeling, P. J. & McCutcheon, J. P. (2017) Endosymbiosis: the feeling is not mutual. *Journal of Theoretical Biology*, 434, 75–79.

15. Sachs, J. L. et al. (2014) Evolutionary origins and diversification of proteobacterial mutualists. *Proceedings of the Royal Society B*, 281, 20132146. https://doi.org/10.1098/rspb.2013.2146.

16. Kaiser, W. et al. (2010) Plant green-island phenotype induced by leaf-miners is mediated by bacterial symbionts. *Proceedings of the Royal Society B*, 277, 2311–2319.

17. Van Baalen, M. & Jansen, V. A. A. (2001) Dangerous liaisons: the ecology of private interest and common good. *Oikos*, 95, 211–224; Johnson, C. A. et al. (2021) Coevolutionary transitions from antagonism to mutualism explained by the co-opted antagonist hypothesis. *Nature Communications*, 12. https://doi.org/10.1038/s41467-021-23177-x.

18. Jeon, K. W. (1972) Development of cellular dependence on infective organisms: micrurgical studies in amoebas. *Science*, 176, 1122–1123; Jeon, K. W. (1987) Change of cellular pathogens into required cell components. *Annals of the New York Academy of Sciences*, 503, 359–371; Jeon, T. J. & Jeon, K. W. (2004) Gene switching in *Amoeba proteus* caused by endosymbiotic bacteria. *Journal of Cell Science*, 117, 535–543.

19. Weeks, A. R. et al. (2007) From parasite to mutualist: rapid evolution of Wolbachia in natural populations of Drosophila. *PloS Biol*, 5, e114. https://doi.org/10.1371/journal.pbio.0050114.

20. Correa, C. C. & Ballard, J. W. O. (2016) *Wolbachia* associations with insects: winning or losing against a master manipulator. *Frontiers in Ecology and Evolution*, 3. https://doi.org/10.3389/fevo.2015.00153.

21. McCutcheon, J. P., Boyd, B. M., & Dale, C. (2019) the life of an insect endosymbiont from the cradle to the grave. *Current Biology*, 29, R485–R495.

22. Matsuura, Y. et al. (2018) Recurrent symbiont recruitment from fungal parasites in cicadas. *proceedings of the National Academy of Sciences,* 115, E5970–E5979; Haji, D. et al. (2021) Host-associated microbial diversity in New Zealand cicadas uncovers elevational structure and replacement of obligate bacterial endosymbionts by *Ophiocordyceps* fungal pathogens. *bioRxiv.* https://doi. org/10.1101/2021.08.24.457591.

CHAPTER 7

1. UFO Sightings Map: https://www.arcgis.com/apps/webappviewer/index. html?id=ddda71d5211f47e782b12f3f8d06246e [Accessed 8/12/20].
2. Garwood, R. J., Oliver, H., & Spencer, A. R. T. (2020) An introduction to the Rhynie chert. *Geological Magazine,* 157, 47–64; Strullu-Derrien, C., Kenrick, P., & Knoll, A. H. (2019) The Rhynie chert. *Current Biology,* 29, R1218–R1223.
3. Hoysted, G. A. et al. (2018) A mycorrhizal revolution. *Curr Opin Plant Biol,* 44, 1–6.
4. Delavaux, C. S., Smith-Ramesh, L. M., & Kuebbing, S. E. (2017) Beyond nutrients: a meta-analysis of the diverse effects of arbuscular mycorrhizal fungi on plants and soils. *Ecology,* https://doi.org/10.1002/ecy.1892.
5. Smith, S. E. & Read, D. (2008) The symbionts forming arbuscular mycorrhizas. In *Mycorrhizal Symbiosis,* 3rd edition (eds S. E. Smith & D. Read). Academic Press, London, pp. 13–41.
6. Corradi, N. & Brachmann, A. (2017) Fungal mating in the most widespread plant symbionts? *Trends in Plant Science,* 22, 175–183.
7. Pawlowska, T. E. et al. (2018) Biology of fungi and their bacterial endosymbionts. *Annual Review of Phytopathology,* 56, 289–309; Mondo, S. J. (2012) Evolutionary stability in a 400-million-year-old heritable facultative mutualism. *Evolution,* 66, 2564–2576.
8. Bever, J. D. (2015) Preferential allocation, physio-evolutionary feedbacks, and the stability and environmental patterns of mutualism between plants and their root symbionts. *New Phytologist,* 205, 1503–1514.
9. Kiers, E. T. et al. (2011) Reciprocal rewards stabilize cooperation in the mycorrhizal symbiosis. *Science,* 333, 880–882; Wyatt, G. A. K. et al. (2014) A biological market analysis of the plant-mycorrhizal symbiosis. *Evolution,* 68, 2603–2618.
10. Perez-Lamarque, B. et al. (2020) Cheating in arbuscular mycorrhizal mutualism: a network and phylogenetic analysis of mycoheterotrophy. *New Phytologist,* 226, 1822–1835; Gebauer, G., Preiss, K., & Gebauer, A. C. (2016) Partial mycoheterotrophy is more widespread among orchids than previously assumed. *New Phytologist,* 211, 11–15.
11. http://www.theplantlist.org/1.1/browse/A/Orchidaceae/ [Accessed 20.02.22].
12. Pecoraro, L. et al. (2018) Fungal networks and orchid distribution: new insights from above- and below-ground analyses of fungal communities. *Ima Fungus,* 9, 1–11.

13. Jacquemyn, H. & Merckx, V. (2019) Mycorrhizal symbioses and the evolution of trophic modes in plants. *Journal of Ecology*, 107, 1567–1581.
14. Suetsugu, K., Haraguchi, T. F., & Tayasu, I. (2022) Novel mycorrhizal cheating in a green orchid: *Cremastra appendiculata* depends on carbon from deadwood through fungal associations. *New Phytologist*. https://doi.org/10.1111/nph.17313.
15. Radhakrishnan, G. V. (2020) An ancestral signalling pathway is conserved in intracellular symbioses-forming plant lineages. *Nature Plants*, 6, 280–289.
16. Rey, T. et al. The *Medicago truncatula* GRAS protein RAD1 supports arbuscular mycorrhiza symbiosis and *Phytophthora palmivora* susceptibility. *Journal of Experimental Botany*, 68, 5871–5881 (2017).
17. Miyauchi, S. et al. (2020) Large-scale genome sequencing of mycorrhizal fungi provides insights into the early evolution of symbiotic traits. *Nature Communications*, 11. https://doi.org/10.1038/s41467-020-18795-w.
18. Alvarez, W. (1998) *T. rex and the Crater of Doom*. Penguin, London.
19. Werner, G. D. et al. (2014) A single evolutionary innovation drives the deep evolution of symbiotic N2-fixation in angiosperms. *Nature Communication*, 5, 4087. https://doi.org/10.1038/ncomms5087.
20. Herridge, D. F., Peoples, M. B., & Boddey, R. M. (2008) Global inputs of biological nitrogen fixation in agricultural systems. *Plant and Soil*, 311, 1–18.
21. Capdevila-Cortada, M. (2019) Electrifying the Haber–Bosch. *Nature Catalysis*, 2, 1055–1055.
22. Santana, M. A. et al. (1998) Evidence that the plant host synthesizes the heme moiety of leghemoglobin in root nodules. *Plant physiology*, 116, 1259–1269.
23. Impossible_burgers (2022) What is soy leghemoglobin, or heme? https://faq.impossiblefoods.com/hc/en-us/articles/360019100553-What-is-soy-leghemoglobin-or-heme- [Accessed 22.02.22].
24. Wardell, G. E. et al. (2022) Why are rhizobial symbiosis genes mobile? *Philosophical Transactions of the Royal Society B: Biological Sciences*, 377. https://doi.org/10.1098/rstb.2020.0471.
25. Pi, H. W. et al. (2022) Origin and evolution of nitrogen fixation in prokaryotes. *Molecular Biology and Evolution*, 39. https://doi.org/10.1093/molbev/msac181.
26. Griesmann, M. et al. (2018) Phylogenomics reveals multiple losses of nitrogen-fixing root nodule symbiosis. *Science*, 361, 1–11.
27. Silvertown, J. (2005) *Demons in Eden: The Paradox of Plant Diversity*. Chicago University Press, Chicago.
28. Simard, S. (2021) *Finding the Mother Tree: Discovering the Wisdom and Intelligence of the Forest*. Allen Lane, London.
29. Tedersoo, L., Bahram, M., & Zobel, M. (2020) How mycorrhizal associations drive plant population and community biology. *Science*, 367, 867. https://doi.org/10.1126/science.aba1223; Klein, T., Siegwolf, R. T. W., & Körner, C. (2016) Belowground carbon trade among tall trees in a temperate forest. *Science*, 352, 342–344.
30. Karst, J., Jones, M. D., & Hoeksema, J. D. (2023) Positive citation bias and overinterpreted results lead to misinformation on common mycorrhizal networks in forests. *Nature Ecology & Evolution*. https://doi.org/10.1038/s41559-023-01986-1.

31. Robinson, D. & Fitter, A. (1999) The magnitude and control of carbon transfer between plants linked by a common mycorrhizal network. *Journal of Experimental Botany*, 50, 9–13.

32. Fraser, E. C., Lieffers, V. J., & Landhausser, S. M. (2006) Carbohydrate transfer through root grafts to support shaded trees. *Tree Physiology*, 26, 1019–1023.

33. Wohlleben, P. & Flannery, T. (2016) *The Hidden Life of Trees: What They Feel, How They Communicate—Discoveries from a Secret World*. Greystone Books, Vancouver.

34. Hafner, B. D., Hesse, B. D., & Grams, T. E. E. (2021) Friendly neighbours: hydraulic redistribution accounts for one quarter of water used by neighbouring drought stressed tree saplings. *Plant Cell and Environment*, 44, 1243–1256.

35. Caldwell, M. M. & Richards, J. H. (1989) Hydraulic lift: water efflux from upper roots improves effectiveness of water-uptake by deep roots. *Oecologia*, 79, 1–5.

36. Callaway, R. M. et al. (2002) Positive interactions among alpine plants increase with stress. *Nature*, 417, 844–848.

CHAPTER 8

1. Bergstrom, A. et al. (2022) Grey wolf genomic history reveals a dual ancestry of dogs. *Nature*, 607, 313–320.

2. The Empress of Blandings is Lord Emsworth's prize pig in novels and short stories by P. G. Wodehouse. Wodehouse, P. G. (1976) *The World of Blandings*. Barrie and Jenkins, London.

3. Juge, A. E., Foster, M. F., & Daigle, C. L. (2022) Canine olfaction as a disease detection technology: a systematic review. *Applied Animal Behaviour Science*, 253. https://doi.org/10.1016/j.applanim.2022.105664.

4. Francis, R. C. (2015) *Domesticated: Evolution in a Man-Made World*. W.W. Norton & Company, New York.

5. Pettitt, P. & Bahn, P. (2014) Against Chauvet-nism: a critique of recent attempts to validate an early chronology for the art of Chauvet Cave. *L'Anthropologie*, 118, 163–182.

6. Sykes, B. (2019) *The Wolf Within: The Astonishing Evolution of the Wolf into Man's Best Friend*. William Collins, London.

7. Bergström, A. et al. (2020) Origins and genetic legacy of prehistoric dogs. *Science*, 370, 557–564.

8. Silvertown, J. W. (2017) *Dinner with Darwin: Food, Drink, and Evolution*. The University of Chicago Press, Chicago.

9. Bryce, C. M. (2021) Dogs as pets and pests: global patterns of canine abundance, activity, and health. *Integrative and Comparative Biology*, 61, 154–165.

10. Sykes, B. (2019) *The Wolf Within: The Astonishing Evolution of the Wolf into Man's Best Friend*. William Collins, London, p. 86.

11. Francis, R. C. (2015) *Domesticated: Evolution in a Man-Made World*. W.W. Norton & Company, New York, p. 32.

12. Columella *De re Rustica Vol. II Book 7*. Translated by E. S. Forster & E. H. Heffner (1954), Harvard University Press, Cambridge, MA. As quoted in Sykes, B. (2019)

The Wolf Within: The Astonishing Evolution of the Wolf into Man's Best Friend. William Collins, London.

13. Paul, M. & Bhadra, A. (2018) The great Indian joint families of free-ranging dogs. *PLoS One,* 13, e0197328. https://doi.org/10.1371/journal.pone.0197328.

14. https://www.thekennelclub.org.uk/breed-standards/gundog/retriever-golden/ [Accessed 11.08.22].

15. Wayne, R. K. & vonHoldt, B. M. (2012) Evolutionary genomics of dog domestication. *Mamm Genome,* 23, 3–18.

16. Driscoll, C. A. et al. (2007) The Near Eastern origin of cat domestication. *Science,* 317, 519–523.

17. Cucchi, T. et al. (2020) Tracking the Near Eastern origins and European dispersal of the western house mouse. *Scientific Reports,* 10. https://doi.org/10.1038/s41598-020-64939-9; Weissbrod, L. et al. (2017) Origins of house mice in ecological niches created by settled hunter-gatherers in the Levant 15,000 y ago. *Proceedings of the National Academy of Sciences,* 114, 4099–4104.

18. Jones, E. P. et al. (2013) Genetic tracking of mice and other bioproxies to infer human history. *Trends Genet,* 29, 298–308.

19. Ottoni, C. & Van Neer, W. V. (2020) The dispersal of the domestic cat paleogenetic and zooarcheological evidence. *Near Eastern Archaeology,* 83, 38–45.

20. Bradshaw, J. W. S. (2016) Sociality in cats: a comparative review. *Journal of Veterinary Behavior: Clinical Applications and Research,* 11, 113–124.

21. Driscoll, C. A., Macdonald, D. W., & O'Brien, S. J. (2009) From wild animals to domestic pets: an evolutionary view of domestication. *Proceedings of the National Academy of Sciences,* 106, 9971–9978.

22. Montague, M. J. et al. (2014) Comparative analysis of the domestic cat genome reveals genetic signatures underlying feline biology and domestication. *Proceedings of the National Academy of Sciences.* https://doi.org/10.1073/pnas.1410083111.

23. Jardat, P. & Lansade, L. (2022) Cognition and the human–animal relationship: a review of the sociocognitive skills of domestic mammals toward humans. *Animal Cognition,* 25, 369–384.

24. Kaminski, J. et al. (2019) Evolution of facial muscle anatomy in dogs. *Proceedings of the National Academy of Sciences,* 116, 14677–14681.

25. Diamond, J. (1997) *Guns, Germs and Steel.* Chatto & Windus, London.

26. Berdoy, M., Webster, J. P., & Macdonald, D. W. (2000) Fatal attraction in rats infected with Toxoplasma gondii. *Proceedings of the Royal Society B: Biological Sciences,* 267, 1591–1594.

27. Gering, E. et al. (2021) Toxoplasma gondii infections are associated with costly boldness toward felids in a wild host. *Nature Communications,* 12. https://doi.org/10.1038/s41467-021-24092-x.

28. Johnson, H. J. & Koshy, A. A. (2020) Latent toxoplasmosis effects on rodents and humans: how much is real and how much is media hype? *Mbio,* 11. https://doi.org/10.1128/mBio.02164-19; Sutterland, A. L. et al. (2015) Beyond the association. *Toxoplasma gondii* in schizophrenia, bipolar disorder, and addiction—systematic review and meta-analysis. *Acta Psychiatrica Scandinavica,* 132, 161–179.

29. Lucking, R. et al. (2009) Do lichens domesticate photobionts like farmers domesticate crops? Evidence from a previously unrecognized lineage of filamentous cyanobacteria. *American Journal of Botany*, 96, 1409–1418.
30. Schultz, T. R. & Brady, S. G. (2008) Major evolutionary transitions in ant agriculture. *Proceedings of the National Academy of Sciences*, 105, 5435–5440.
31. Chapela, I. H., Rehner, S. A., Schultz, T. R., & Mueller, U. G. (1994) Evolutionary history of the symbiosis between fungus-growing ants and their fungi. *Science*, 266, 1691–1694.
32. Mehdiabadi, N. J., Hughes, B., & Mueller, U. G. (2006) Cooperation, conflict, and coevolution in the attine ant-fungus symbiosis. *Behavioral Ecology*, 17, 291–296.
33. Darwin, C. (1868) *The Variation of Animals and Plants under Domestication*. John Murray, London.
34. Wilkins, A. S., Wrangham, R. W., & Fitch, W. T. (2014) The 'domestication syndrome' in mammals: a unified explanation based on neural crest cell behavior and genetics. *Genetics*, 197, 795–808.
35. Rubio, A. O. & Summers, K. (2022) Neural crest cell genes and the domestication syndrome: A comparative analysis of selection. *Plos One*, 17. https://doi.org/10.1371/journal.pone.0263830.
36. Bagehot, W. (1872) *Physics and Politics, Or, Thoughts on the Application of the Principles of 'natural Selection' and "inheritance" to Political Science*. Henry S. King & Co., London.
37. Darwin, C. (1871) *The Descent of Man and Selection in Relation to Sex*. J. Murray, London, chap. 4.
38. Burkart, J. M. et al. (2014) The evolutionary origin of human hypercooperation. *Nature Communications*, 5. https://doi.org/10.1038/ncomms5747.
39. Henrich, J. (2015) *The Secret of Our Success: How Culture Is Driving Human Evolution, Domesticating Our Species, and Making Us Smarter*. Princeton University Press, Princeton, NJ.
40. Hare, B. (2017) Survival of the friendliest: *Homo sapiens* evolved via selection for prosociality. *Annual Review of Psychology*, 68, 155–186 (ed. S. T. Fiske).
41. Theofanopoulou, C. et al. (2017) Self-domestication in Homo sapiens: insights from comparative genomics. *PloS one*, 12, e0185306–e0185306.
42. Sánchez-Villagra, M. R. & van Schaik, C. P. (2019) Evaluating the self-domestication hypothesis of human evolution. *Evolutionary Anthropology*, 28, 133–143.
43. Herron, M. D. (2021) What are the major transitions? *Biology & Philosophy*, 36. https://doi.org/10.1007/s10539-020-09773-z.
44. Boomsma, J. J. (2023) *Domains and Major Transitions of Social Evolution*. Oxford University Press, Oxford.

CHAPTER 9

1. Giuffrida, A. (2022) Florence tomb by Michelangelo restored with aid of bacteria. *The Guardian*, 9 March.

2. Rul, F. & Monnet, V. (2015) How microbes communicate in food: a review of signalling molecules and their impact on food quality. *Current Opinion in Food Science*, 2, 100–105.

3. Simonet, C. & McNally, L. (2021) Kin selection explains the evolution of cooperation in the gut microbiota. *Proceedings of the National Academy of Sciences*, 118. https://doi.org/10.1073/pnas.2016046118; Smith, D. (1956) *The One Hundred and One Dalmatians*. Heinemann, London.

4. Flemming, H. C. & Wuertz, S. (2019) Bacteria and archaea on Earth and their abundance in biofilms. *Nature Reviews Microbiology*, 17, 247–260.

5. Wucher, B. R. et al. (2021) Bacterial predation transforms the landscape and community assembly of biofilms. *Current Biology*, 31, 2643–2651.

6. Nadell, C. D., Xavier, J. B., & Foster, K. R. (2009) The sociobiology of biofilms. *Fems Microbiology Reviews*, 33, 206–224.

7. Zarubin, M. et al. (2012) Bacterial bioluminescence as a lure for marine zooplankton and fish. *Proceedings of the National Academy of Sciences*, 109, 853–7.

8. Riiser, E. S. et al. (2019) Switching on the light: using metagenomic shotgun sequencing to characterize the intestinal microbiome of Atlantic cod. *Environmental Microbiology*, 21, 2576–2594.

9. Bruger, E. L. et al. (2021) Quorum sensing provides a molecular mechanism for evolution to tune and maintain investment in cooperation. *ISME Journal*, 15, 1236–1247.

10. Popat, R. et al. (2012) Quorum-sensing and cheating in bacterial biofilms. *Proceedings of the Royal Society B*, 279, 4765–4771.

11. Sikdar, R. & Elias, M. (2020) Quorum quenching enzymes and their effects on virulence, biofilm, and microbiomes: a review of recent advances. *Expert Review of Anti-infective Therapy*, 18, 1221–1233.

12. Galan, J. E. & Waksman, G. (2018) Protein-injection machines in bacteria. *Cell*, 172, 1306–1318.

13. Majerczyk, C., Schneider, E., & Greenberg, E. P. (2016) Quorum sensing control of Type VI secretion factors restricts the proliferation of quorum-sensing mutants. *Elife*, 5, e14712. https://doi.org/10.7554/eLife.14712.

14. Kehe, J. et al. (2021) Positive interactions are common among culturable bacteria. *Science Advances*, 7, eabi7159. https://doi.org/10.1126/sciadv.abi7159.

15. Palmer, J. D. & Foster, K. R. (2022) Bacterial species rarely work together. *Science*, 376, 581–582.

16. Lagkouvardos, I., Overmann, J., & Clavel, T. (2017) Cultured microbes represent a substantial fraction of the human and mouse gut microbiota. *Gut Microbes*, 8, 493–503.

17. Thompson, L. R. et al. (2017) A communal catalogue reveals Earth's multiscale microbial diversity. *Nature*, 551, 457–463.

18. Machado, D. et al. (2021) Polarization of microbial communities between competitive and cooperative metabolism. *Nature Ecology & Evolution*. https://doi.org/10.1038/s41559-020-01353-4.

19. D'Souza, G. et al. (2018) Ecology and evolution of metabolic cross-feeding interactions in bacteria. *Natural Product Reports*, 35, 455–488.

20. Rakoff-Nahoum, S., Foster, K. R., & Comstock, L. E. (2016) The evolution of cooperation within the gut microbiota. *Nature*, 533, 255–259.

21. Oliveira, N. M., Niehus, R., & Foster, K. R. (2014) Evolutionary limits to cooperation in microbial communities. *Proceedings of the National Academy of Sciences*, 111, 17941–17946.

22. Shkoporov, A. N., Turkington, C. J., & Hill, C. (2022) Mutualistic interplay between bacteriophages and bacteria in the human gut. *Nature Reviews Microbiology*. https://doi.org/10.1038/s41579-022-00755-4.

23. Breitbart, M. (2012) Marine viruses: truth or dare. *Annual Review of Marine Science*, 4, 425–448 (eds C. A. Carlson & S. J. Giovannoni).

24. Morris, J. J., Lenski, R. E., & Zinser, E. R. (2012) The black queen hypothesis: evolution of dependencies through adaptive gene loss. *mBio*, 3, e00036-12. https://doi.org/10.1128/mBio.00036-12; D'Souza, G. et al. (2014) Less is more: selective advantages can explain the prevalent loss of biosynthetic genes in bacteria. *Evolution*, 68, 2559–2570.

25. Morris, J. J. et al. (2011) Dependence of the cyanobacterium Prochlorococcus on hydrogen peroxide scavenging microbes for growth at the ocean's surface. *PLoS One*, 6, e16805. https://doi.org/10.1371/journal.pone.0016805.

26. Piccardi, P., Vessman, B., & Mitri, S. (2019) Toxicity drives facilitation between 4 bacterial species. *Proceedings of the National Academy of Sciences*, 116, 15979–15984.

27. Hesse, E. et al. (2021) Stress causes interspecific facilitation within a compost community. *Ecology Letters*, 24, 2169–2177. 1

28. Huang, X., Xin, Y., & Lu, T. (2022) A systematic, complexity-reduction approach to dissect the kombucha tea microbiome. *Elife*, 11. https://doi.org/10.7554/eLife.76401.

29. Celiker, H. & Gore, J. (2012) Competition between species can stabilize public-goods cooperation within a species. *Molecular Systems Biology*, 8. https://doi.org/10.1038/msb.2012.54.

30. Greig, D. & Travisano, M. (2004) The prisoner's dilemma and polymorphism in yeast SUC genes. *Proceedings of the Royal Society B: Biological Sciences*, 271, S25–S26.

31. Maclean, R. C. & Brandon, C. (2008) Stable public goods cooperation and dynamic social interactions in yeast. *Journal of Evolutionary Biology*, 21, 1836–1843.

32. Gardner, A. & West, S. A. (2010) Greenbeards. *Evolution*, 64, 25–38.

33. Smukalla, S. et al. (2008) FLO1 is a variable green beard gene that drives biofilm-like cooperation in budding yeast. *Cell*, 135, 726–737.

CHAPTER 10

1. Sapp, J. (2005) The prokaryote-eukaryote dichotomy: meanings and mythology. *Microbiology and Molecular Biology Reviews*, 69, 292–305.

2. Balch, W. E., Magrum, L. J., Fox, G. E., Wolfe, R. S., & Woese, C. R. (1977) An ancient divergence among the bacteria. *Journal of Molecular Evolution*, 9, 305–311.

3. Sapp, J. (2005) The prokaryote-eukaryote dichotomy: meanings and mythology. *Microbiology and Molecular Biology Reviews*, 69, 292–305.

4. Marin, J., Battistuzzi, F. U., Brown, A. C., & Hedges, S. B. (2017) The timetree of prokaryotes: new insights into their evolution and speciation. *Molecular Biology and Evolution*, 34, 437–446.

5. It was thought until recently that only archaea produced methane, but recent research shows that cyanobacteria can also do this. Bizic, M. et al. (2020) Aquatic and terrestrial cyanobacteria produce methane. *Science Advances*, 6. https://doi.org/10.1126/sciadv.aax5343.

6. Bryant, M. P., Wolin, E. A., Wolin, M. J., & Wolfe, R. S. (1967) *Methanobacillus omelianskii*, a symbiotic association of two species of bacteria. *Archives of Microbiology*, 59, 20–31.

7. Eme, L., Spang, A., Lombard, J., Stairs, C. W., & Ettema, T. J. G. (2017) Archaea and the origin of eukaryotes. *Nature Reviews Microbiology*, 15, 711–723.

8. Betts, H. C. et al. (2018) Integrated genomic and fossil evidence illuminates life's early evolution and eukaryote origin. *Nature Ecology & Evolution*, 2, 1556–1562.

9. Spang, A. et al. (2015) Complex archaea that bridge the gap between prokaryotes and eukaryotes. *Nature*, 521, 173–178.

10. Spang, A. et al. (2015) Complex archaea that bridge the gap between prokaryotes and eukaryotes. *Nature*, 521, 173–178.

11. Williams, T. A. et al. (2020) Phylogenomics provides robust support for a two-domains tree of life. *Nature Ecology & Evolution*, 4, 138–147.

12. Imachi, H. et al. (2020) Isolation of an archaeon at the prokaryote–eukaryote interface. *Nature*. https://doi.org/10.1038/s41586-019-1916-6.

13. Rodrigues-Oliveira, T., Wollweber, F., Ponce-Toledo, R. I., Xu, J., Rittmann, S. K. R., Klingl, A., Pilhofer, M., & Schleper, C. (2022) Actin cytoskeleton and complex cell architecture in an Asgard archaeon. *Nature*. https://doi.org/10.1038/s41586-022-05550-y.

14. Lopez-Garcia, P. & Moreira, D. (2020) The syntrophy hypothesis for the origin of eukaryotes revisited. *Nature Microbiology*, 5, 655–667.

15. Baum, D. A. & Baum, B. (2014) An inside-out origin for the eukaryotic cell. *BMC Biology*, 12. https://doi.org/10.1186/s12915-014-0076-2.

16. Gould, S. B. (2018) Membranes and evolution. *Current Biology*, 28, R381–R385.

17. Rivera, M. C., Jain, R., Moore, J. E., & Lake, J. A. (1998) Genomic evidence for two functionally distinct gene classes. *Proceedings of the National Academy of Sciences*, 95, 6239–6244.

18. Wein, T. & Sorek, R. (2022) Bacterial origins of human cell-autonomous innate immune mechanisms. *Nature Reviews Immunology*. https://doi.org/10.1038/s41577-022-00705-4.

19. Brueckner, J. & Martin, W. F. (2020) Bacterial genes outnumber archaeal genes in eukaryotic genomes. *Genome Biology and Evolution*, 12, 282–292.

20. Gabaldon, T. (2021) Origin and early evolution of the eukaryotic cell. *Annual Review of Microbiology*, 75, 631–647 (ed. S. Gottesman).

21. Schön, M. E. et al. (2022) The evolutionary origin of host association in the Rickettsiales. *Nature Microbiology*, 7, 1189–1199.

22. Zachar, I. & Szathmary, E. (2017) Breath-giving cooperation: critical review of origin of mitochondria hypotheses Major unanswered questions point to the importance of early ecology. *Biology Direct*, 12. https://doi.org/10.1186/s13062-017-0190-5.

23. Sagan, L. (1967) On the origin of mitosing cells. *Journal of Theoretical Biology*, 14, 255–274.

24. Martin, W. F. (2017) Physiology, anaerobes, and the origin of mitosing cells 50 years on. *Journal of Theoretical Biology*, 434, 2–10.

25. Shiflett, A. M. & Johnson, P. J. (2010) Mitochondrion-related organelles in eukaryotic protists. *Annual Review of Microbiology*, 64, 409–429.

26. Gumsley, A. P. et al. (2017) Timing and tempo of the Great Oxidation Event. *Proceedings of the National Academy of Sciences*, 114, 1811–1816.

27. Mills, D. B. et al. (2022) Eukaryogenesis and oxygen in Earth history. *Nature Ecology & Evolution*, 6, 520–532.

28. Martin, W. F., Mentel, M., & Tielens, A. G. M. (2020) *Mitochondria and Anaerobic Energy Metabolism in Eukaryotes: Biochemistry and Evolution*. De Gruyter, Berlin.

29. Cohen, P. A. & Kodner, R. B. (2022) The earliest history of eukaryotic life: uncovering an evolutionary story through the integration of biological and geological data. *Trends in Ecology & Evolution*, 37, 246–256; Mills, D. B. et al. (2022) Eukaryogenesis and oxygen in Earth history. *Nature Ecology & Evolution*, 6, 520–532.

30. Speijer, D. (2022) Eukaryotes were shaped by Oxygen. *Nature Ecology and Evolution*, 6, 1242.

31. Horn, R., Gupta, K. J., & Colombo, N. (2014) Mitochondrion role in molecular basis of cytoplasmic male sterility. *Mitochondrion*, 19 Pt B, 198–205.

32. Dornier, A. & Dufay, M. (2013) How selfing, inbreeding depression, and pollen limitation impact nuclear-cytoplasmic gynodioecy: A model. *Evolution*, 67, 2674–2687.

33. Havird, J. C. et al. (2019) Selfish mitonuclear conflict. *Current Biology*, 29, R496–R511.

34. Any similarity with the effects of Brexit on the British economy is merely fortuitous.

35. Roger, A. J., Susko, E., & Leger, M. M. (2021) Evolution: reconstructing the timeline of eukaryogenesis. *Current Biology*, 31, R193–R196.

CHAPTER 11

1. Jablonska, J. & Tawfik, D. S. (2021) The evolution of oxygen-utilizing enzymes suggests early biosphere oxygenation. *Nature Ecology & Evolution*, 5, 442–448.

2. Oliver, T. et al. (2021) Time-resolved comparative molecular evolution of oxygenic photosynthesis. *Biochimica Et Biophysica Acta-Bioenergetics*, 1862. https://doi.org/10.1016/j.bbabio.2021.148400.

3. Soo, R. M. et al. (2017) On the origins of oxygenic photosynthesis and aerobic respiration in Cyanobacteria. *Science*, 355, 1436–1439.

4. Raymond, J. et al. (2002) Whole-genome analysis of photosynthetic prokaryotes. *Science*, 298, 1616–1620.
5. Sanchez-Baracaldo, P. & Cardona, T. (2019) On the origin of oxygenic photosynthesis and Cyanobacteria. *New Phytologist*. https://doi.org/10.1111/nph.16249.
6. There is one recently discovered exception called *Paulinella* that is mentioned later.
7. Gibson, T. M. et al. (2018) Precise age of *Bangiomorpha pubescens* dates the origin of eukaryotic photosynthesis. *Geology*, 46, 135–138.
8. Nowack, E. C. M. & Weber, A. P. M. (2018) Genomics-informed insights into endosymbiotic organelle evolution in photosynthetic eukaryotes. *Annual Review of Plant Biology*, 69, 51–84 (ed. S. S. Merchant).
9. Gavelis, G. S. & Gile, G. H. (2018) How did cyanobacteria first embark on the path to becoming plastids? Lessons from protist symbioses. *Fems Microbiology Letters*, 365. https://doi.org/10.1093/femsle/fny209.
10. Sanchez-Baracaldo, P., Raven, J. A., Pisani, D., & Knoll, A. H. (2017) Early photosynthetic eukaryotes inhabited low-salinity habitats. *Proceedings of the National Academy of Sciences*, 114, E7737–E7745.
11. Sanders, W. B. (2022) The photoaerogens: algae and plants reunited conceptually. *American Journal of Botany*, 109, 363–365.
12. Dyer, B. D. (2003) *A Field Guide to Bacteria*. Cornell University Press, Ithaca, NY.
13. Bunker, F. StP. et al. (2017) *Seaweeds of Britain and Ireland*. Wild Nature Press, Plympton St Maurice, Plymouth.
14. Strassert, J. F. H., Irisarri, I., Williams, T. A., & Burki, F. (2021) A molecular timescale for eukaryote evolution with implications for the origin of red algal-derived plastids. *Nature Communications*, 12. https://doi.org/10.1038/s41467-021-22044-z.
15. Macorano, L. & E. C. M. Nowack (2021) *Paulinella chromatophora*. *Current Biology*, 31, R1024–R1026.
16. Schuster, F. L. (2002) Cultivation of pathogenic and opportunistic free-living amebas. *Clinical Microbiology Reviews*, 15, 342–354.
17. Stephens, T. G. et al. (2021) Why is primary endosymbiosis so rare? *New Phytologist*, 231, 1693–1699.
18. Sibbald, S. J. & Archibald, J. M. (2020) Genomic insights into plastid evolution. *Genome Biology and Evolution*, 12, 978–990.
19. Waller, R. F. & Koreny, L. (2017) Plastid complexity in dinoflagellates: a picture of gains, losses, replacements and revisions. In *Secondary Endosymbioses*. Elsevier, Amsterdam (ed. Y. Hirakawa), pp. 105–143.
20. Hadariova, L. et al. (2018) Reductive evolution of chloroplasts in non-photosynthetic plants, algae and protists. *Current Genetics*, 64, 365–387.
21. Salomaki, E. D. & Lane, C. E. (2014) Are all red algal parasites cut from the same cloth? *Acta Societatis Botanicorum Poloniae*, 83, 369–375.
22. Janouskovec, J. et al. (2010) A common red algal origin of the apicomplexan, dinoflagellate, and heterokont plastids. *Proceedings of the National Academy of Sciences*, 107, 10949–10954.

23. Biddau, M. & Sheiner, L. (2019) Targeting the apicoplast in malaria. *Biochemical Society Transactions*, 47, 973–983.
24. Martin, W. et al. (2002) Evolutionary analysis of *Arabidopsis*, cyanobacterial, and chloroplast genomes reveals plastid phylogeny and thousands of cyanobacterial genes in the nucleus. *Proceedings of the National Academy of Sciences*, 99, 12246–12251.

CHAPTER 12

1. Burki, F. et al. (2020) The new tree of eukaryotes. *Trends in Ecology & Evolution*, 35, 43–55.
2. Boraas, M. E., Seale, D. B., & Boxhorn, J. E. (1998) Phagotrophy by a flagellate selects for colonial prey: a possible origin of multicellularity. *Evolutionary Ecology*, 12, 153–164.
3. Fisher, R. M., Bell, T., & West, S. A. (2016) Multicellular group formation in response to predators in the alga *Chlorella vulgaris*. *Journal of Evolutionary Biology*, 29, 551–559; Kapsetaki, S. E. & West, S. A. (2019) The costs and benefits of multicellular group formation in algae. *Evolution*, 73, 1296–1308.
4. Grosberg, R. K. & Strathmann, R. R. (2007) The evolution of multicellularity: a minor major transition? *Annual Review of Ecology Evolution and Systematics*, 38, 621–654.
5. Umen, J. & Herron, M. D. (2021) Green algal models for multicellularity. *Annual Review of Genetics*, 55, 603–632.
6. Konig, S. G. & Nedelcu, A. M. (2020) The genetic basis for the evolution of soma: mechanistic evidence for the co-option of a stress-induced gene into a developmental master regulator. *Proceedings of the Royal Society B-Biological Sciences*, 287. https://doi.org/10.1098/rspb.2020.1414.
7. Joint, I., Tait, K., & Wheeler, G. (2007) Cross-kingdom signalling: exploitation of bacterial quorum sensing molecules by the green seaweed *Ulva*. *Philosophical Transactions of the Royal Society B: Biological Sciences*, 362, 1223–1233.
8. Grosberg, R. K. & Strathmann, R. R. (1998) One cell, two cell, red cell, blue cell: the persistence of a unicellular stage in multicellular life histories. *Trends in Ecology & Evolution*, 13, 112–116.
9. Bonner, J. T. (2008) *The Social Amoebae: The Biology of Cellular Slime Molds*. Princeton University Press, Princeton.
10. Cell length of *Dictyostelidium discoideum* https://bionumbers.hms.harvard.edu/search.aspx?task=searchbytrmorg&log=y&trm=Amoeba%20Dictyostelium%20discoideum&pi=4 [Accessed 23.3.21].
11. Kessin, R. H. (2001) *Dictyostelium: Evolution, Cell Biology, and the Development of Multicellularity*. Cambridge University Press, Cambridge.
12. Medina, J. M. et al. (2019) Cooperation and conflict in the social amoeba *Dictyostelium discoideum*. *International Journal of Developmental Biology*, 63, 371–382.
13. Gilbert, O. M. et al. (2007) High relatedness maintains multicellular cooperation in a social amoeba by controlling cheater mutants. *Proceedings of the National Academy of Sciences*, 104, 8913–8917.

14. Tarnita, C. E., Taubes, C. H., & Nowak, M. A. (2013) Evolutionary construction by staying together and coming together. *Journal of Theoretical Biology*, 320, 10–22.
15. Aktipis, C. A. et al. (2015) Cancer across the tree of life: cooperation and cheating in multicellularity. *Philosophical Transactions of the Royal Society B: Biological Sciences*, 370, 1–21.
16. S. A. Frank and M. A. Nowak, Problems of somatic mutation and cancer, *Bioessays*, 26, 3, 291–299.
17. Mathavarajah, S. et al. (2021) Cancer and the breakdown of multicellularity: what *Dictyostelium discoideum*, a social amoeba, can teach us. *Bioessays*, 43. https://doi.org/10.1002/bies.202000156.
18. Domazet-Lošo, T. & Tautz, D. (2010) Phylostratigraphic tracking of cancer genes suggests a link to the emergence of multicellularity in metazoa. *BMC Biology*, 8, 66–76.
19. Trigos, A. S., Pearson, R. B., Papenfuss, A. T., & Goode, D. L. (2018) How the evolution of multicellularity set the stage for cancer. *British Journal of Cancer*, 118, 145–152.
20. Taddei, M. L., Giannoni, E., Fiaschi, T., & Chiarugi, P. (2012) Anoikis: an emerging hallmark in health and diseases. *Journal of Pathology*, 226, 380–393.
21. Sinha, G. (2022) Tumours can teem with microbes: but what are they doing there? *Science*, 378, 693–694.
22. Claessen, D. et al. (2014) Bacterial solutions to multicellularity: a tale of biofilms, filaments and fruiting bodies. *Nature Reviews Microbiology*, 12, 115–124; Serra, D. O. & Hengge, R. (2021) Bacterial multicellularity: the biology of escherichia coli building large-scale biofilm communities. *Annual Review of Microbiology*, 75, 269–290.
23. Cao, P. B., Dey, a., Vassallo, C. N., & Wall, D. (2015) How myxobacteria cooperate. *Journal of Molecular Biology*, 427, 3709–3721.
24. Lane, N. & Martin, W. F. (2015) Eukaryotes really are special, and mitochondria are why. *Proceedings of the National Academy of Sciences*, 112, E4823–E4823.
25. Grosberg, R. K. & Strathmann, R. R. (2007) The evolution of multicellularity: a minor major transition? *Annual Review of Ecology Evolution and Systematics*, 38, 621–654.

CHAPTER 13

1. Nurse, P. (2020) *What Is Life?* David Fickling Books, Oxford.
2. Higgs, P. G. & Lehman, N. (2015) The RNA world: molecular cooperation at the origins of life. *Nature Reviews Genetics*, 16, 7–17.
3. Huxley, T. H. (1868) On some organisms living at great depths in the North Atlantic ocean. *Quarterly Journal of Microscopical Science*, 8, 203–212.
4. Bailey Jr, H. S. (1972) The background of the Challenger expedition. *American Scientist*, 60, 550–560.
5. Eiseley, L. C. (1958) *The Immense Journey.* Victor Gollancz, London.

6. Linklater, E. (1974) *The Voyage of the Challenger.* Sphere Books, London; Jones, E. (2022) *The Challenger Expedition: Exploring the Ocean's Depths.* Royal Museums Greenwich, Greenwich.

7. Darwin, C. (1880) Sir Wyville Thomson and natural selection. *Nature,* 23, 32.

8. Alberts, B. (2022) *Molecular Biology of the Cell.* W. W. Norton & Company, New York.

9. Bernier, C. R. et al. (2018) Translation: the universal structural core of life. *Molecular Biology and Evolution,* 35, 2065–2076.

10. Bowman, J. C. et al. (2020) Root of the tree: the significance, evolution, and origins of the ribosome. *Chemical Reviews,* 120, 4848–4878.

11. Subramanian, S. (2021) *Dominant Character: The Radical Science and Restless Politics of J. B. S. Haldane.* Atlantic Books, London; Lazcano, A. (2010) Historical development of origins research. *Cold Spring Harbor Perspectives in Biology,* 2, a002089. https://doi.org/10.1101/cshperspect.a002089.

12. Xu, J. et al. (2020) Selective prebiotic formation of RNA pyrimidine and DNA purine nucleosides. *Nature,* 582, 60–66.

13. Bhowmik, S. & Krishnamurthy, R. (2019) The role of sugar-backbone heterogeneity and chimeras in the simultaneous emergence of RNA and DNA. *Nature Chemistry,* 11, 1009–1018.

14. Sutherland, J. D. (2017) Studies on the origin of life: the end of the beginning. *Nature Reviews Chemistry,* 1. https://doi.org/10.1038/s41570-016-0012.

15. Ricardo, A. & Szostak, J. W. (2009) Life on earth. *Scientific American,* 301, 54–61.

16. Szostak, J. W., Bartel, D. P., & Luisi, P. L. (2001) Synthesizing life. *Nature,* 409, 387–390.

17. Levin, S. R. & West, S. A. (2017) The evolution of cooperation in simple molecular replicators. *Proceedings of the Royal Society B,* 284. https://doi.org/10.1098/rspb.2017.1967.

18. National Library of Medicine (2019) Sol Spiegelman: the Sol Spiegelman papers. https://profiles.nlm.nih.gov/spotlight/px/feature/biographical-overview [Accessed 11.12.22].

19. Mills, D. R., Peterson, R. L., & Spiegelman, S. (1967) An extracellular Darwinian experiment with a self-duplicating nucleic acid molecule. *Proceedings of the National Academy of Sciences,* 58, 217–224.

20. Vazquez-Salazar, A. & Chen, I. A. (2022) In vitro evolution: from monsters to mobs. *Current Biology,* 32, R580–R582.

21. Mizuuchi, R., Furubayashi, T., & Ichihashi, N. (2022) Evolutionary transition from a single RNA replicator to a multiple replicator network. *Nature Communications,* 13. https://doi.org/10.1038/s41467-022-29113-x.

22. Haldane, J. B. S. (1929) The origin of life. *Rationalist Annual,* 3, 3–10.

23. Tirard, S. (2017) J. B. S. Haldane and the origin of life. *Journal of Genetics,* 96, 735–739.

24. Dodd, M. S. et al. (2017) Evidence for early life in Earth's oldest hydrothermal vent precipitates. *Nature,* 543, 60–64.

25. Deamer, D. (2021) Where did life begin? Testing ideas in prebiotic analogue conditions. *Life-Basel*, 11. https://doi.org/10.3390/life11020134.
26. Lane, N. (2010) *Life Ascending: The Ten Great Inventions of Evolution*. Profile Books, London.
27. Sutherland, J. D. (2017) Studies on the origin of life: the end of the beginning. *Nature Reviews Chemistry*, 1. https://doi.org/10.1038/s41570-016-0012.
28. Russell, M. J. (2021) The 'water problem' (sic), the illusory pond and life's submarine emergence-a review. *Life-Basel*, 11. https://doi.org/10.3390/life11050429.

CHAPTER 14

1. Allen, G. E. (1978) *Thomas Hunt Morgan: The man and His Science*. Princeton University Press, Princeton, NJ.
2. Maynard Smith, J. & Száthmary, E. (1993) The origin of chromosomes I: selection for linkage. *Journal of Theoretical Biology*, 164, 437–446.
3. Maynard Smith, J. (1979) Hypercycles and the origin of life. *Nature*, 280, 445–446.
4. Campbell, N. A., Urry, L. A., Cain, M. L., Wasserman, S. A., Minorsky, P. V., & Orr, R. B. (2021) *Biology: A Global Approach*. Pearson, New York.
5. Maynard Smith, J. (1978) *The Evolution of Sex*. Cambridge University Press, Cambridge.
6. Ridley, M. (2000) *Mendel's Demon*. Phoenix, London.
7. Queller, D. C. & Strassmann, J. E. (2013) The veil of ignorance can favour biological cooperation. *Biology Letters*, 9. https://doi.org/10.1098/rsbl.2013.0365.
8. Rawls, J. & Kelly, E. (2001) *Justice as Fairness: A Restatement*. Harvard University Press, Cambridge, MA.
9. Ågren, J. A., Haig, D., & McCoy, D. E. (2022) Meiosis solved the problem of gerrymandering. *Journal of Genetics*, 101. https://doi.org/10.1007/s12041-022-01383-w.
10. Veller, C. (2022) Mendel's First Law: partisan interests and the parliament of genes. *Heredity (Edinb)*, 129, 48–55.

CHAPTER 15

1. Zayed, A. A. et al. (2022) Cryptic and abundant marine viruses at the evolutionary origins of Earth's RNA virome. *Science*, 376, 156–162.
2. Wells, J. N. & Feschotte, C. (2020) A field guide to eukaryotic transposable elements. *Annual Review of Genetics*, 54, 539–561.
3. McClintock, B. (1950) The origin and behavior of mutable loci in maize. *Proceedings of the National Academy of Sciences*, 36, 344–355.
4. Leeks, A., West, S. A., & Ghoul, M. (2021) The evolution of cheating in viruses. *Nature Communications*, 12, 6928. https://doi.org/10.1038/s41467-021-27293-6.

5. Burns, K. H. (2020) Our conflict with transposable elements and its implications for human disease. *Annual Review of Pathology: Mechanisms of Disease*, 15, 51–70 (eds A. K. Abbas, J. C. Aster, & M. B. Feany).

6. Roth, G. & Walkowiak, W. (2015) The influence of genome and cell size on brain morphology in amphibians. *Cold Spring Harbor Perspectives in Biology*, 7, a019075. https://doi.org/10.1101/cshperspect.a019075.

7. Lynch, M. & Conery, J. S. (2003) The origins of genome complexity. *Science*, 302, 1401–1404.

8. Slotkin, R. K. & Martienssen, R. (2007) Transposable elements and the epigenetic regulation of the genome. *Nature Reviews Genetics*, 8, 272–285.

9. Rowland, H. M., Saccheri, I. J., & Skelhorn, J. (2022) The peppered moth *Biston betularia*. *Current Biology*, 32, R447–R448.

10. van't Hof, A. E. et al. (2016) The industrial melanism mutation in British peppered moths is a transposable element. *Nature*, 534, 102–105.

11. Volff, J. N. (2006) Turning junk into gold: domestication of transposable elements and the creation of new genes in eukaryotes. *Bioessays*, 28, 913–922.

12. Burt, A. & Trivers, R. (2006) *Genes in Conflict: The Biology of Selfish Genetic Elements*. Belknap Press of Harvard University Press, Cambridge, MA.

13. Liu, S. et al. (2020) Functional regulation of an ancestral RAG transposon ProtoRAG by a trans-acting factor YY1 in lancelet. *Nature Communications*, 11. https://doi.org/10.1038/s41467-020-18261-7.

14. Gilbert, C., Peccoud, J., & Cordaux, R. (2021) Transposable elements and the evolution of insects. *Annual Review of Entomology*, 66, 355–372.

15. Burke, G. R., Hines, H. M., & Sharanowski, B. J. (2021) The presence of ancient core genes reveals endogenization from diverse viral ancestors in parasitoid wasps. *Genome Biology and Evolution*, 13. https://doi.org/10.1093/gbe/evab105.

16. Labusova, J., Konradova, H., & Lipavska, H. (2020) The endangered Saharan cypress (*Cupressus dupreziana*): do not let it get into Charon's boat. *Planta*, 251, 63, 1–13.

17. Hammond, A. et al. (2021) Gene-drive suppression of mosquito populations in large cages as a bridge between lab and field. *Nature Communications*, 12, 45–89; Wang, G. H. et al. (2022) Symbionts and gene drive: two strategies to combat vector-borne disease. *Trends in Genetics*, 38, 708–723.

18. Callies, D. E. (2019) The ethical landscape of gene drive research. *Bioethics*, 33, 1091–1097.

CHAPTER 16

1. Kennedy, P., Uller, T., & Helantera, H. (2014) Are ant supercolonies crucibles of a new major transition in evolution? *Journal of Evolutionary Biology*, 27, 1784–1796.

2. Helantera, H. (2022) Supercolonies of ants (Hymenoptera: Formicidae): ecological patterns, behavioural processes and their implications for social evolution. *Myrmecological News*, 32, 1–22.

3. Van Wilgenburg, E., Torres, C. W., & Tsutsui, N. D. (2010) The global expansion of a single ant supercolony. *Evolutionary Applications*, 3, 136–143.

4. McShea, D. W. (2016) Three trends in the history of life: an evolutionary syndrome. *Evolutionary Biology*, 43, 531–542.

5. Ruse, M. (2013) Wrestling with biological complexity: from Darwin to Dawkins. In *Complexity and the Arrow of Time* (eds C. H. Lineweaver, M. Ruse, & P. C. W. Davies) Cambridge University Press, Cambridge, pp. 279–307.

6. Corning, P. A. & Szathmary, E. (2015) 'Synergistic selection': a Darwinian frame for the evolution of complexity. *Journal of Theoretical Biology*, 371, 45–58.

Glossary

Allele One of a number of alternative versions of a gene.

Altruism A behaviour that benefits another individual at a significant cost to the altruist.

Cheat A non-cooperating individual that deceives a cooperator into providing benefits that are costly and are normally directed to other cooperators.

Chromosome A DNA structure formed by chains of genes and various structural and regulatory proteins. Prokaryote chromosomes are circular. Eukaryote chromosomes are linear and very much more complex and numerous.

Commensal A relationship in which one organism benefits from resources provided by another without harm or benefit to the provider.

Communal breeding Cooperation among unrelated families that raise their offspring together.

Competition A relationship between individuals of the same (intraspecific competition) or different (interspecific competition) species in which each negatively affects the other through the consumption of some resource such as food used by both.

Cooperation A social behaviour in which cooperators confer a benefit on others at a cost to themselves. For cooperation to evolve, there must be direct and/or indirect benefits to the cooperator that balance the costs, as given by **Hamilton's rule**.

Cooperative breeding A behaviour in which members of a family help raise offspring that are not their own. Helpers also breed themselves, which makes cooperative breeding different from eusociality, for which it may be an evolutionary precursor.

Domestication Adaptation in a species to fit the needs of humans, brought about by artificial selection such as in crops or farm animals. Synonymous in animals with tameness. Analogous processes may occur within a species (self-domestication) or between non-human species such as leaf-cutter ants and fungi.

Domestication syndrome A set of traits that appears in domesticated animals as an apparent side effect of artificial selection for tameness. It includes a smaller brain, floppy ears, a curled tail, and a piebald coat. Plants express their own domestication syndrome that for seed crops includes larger seed size, lack of dormancy, and restricted seed dispersal.

Drive (gene drive) Greater than 50 per cent transmission of an allele into the next generation, breaking **gene justice**.

Emergence The process by which interactions among the parts of a system produce non-linear outcomes. Emergent phenomena occur when the whole is greater than the sum of the parts.

Endosymbiont A symbiotic microbe that lives inside the cells of its host. Endosymbionts are often subject to **vertical transmission**.

Eukaryote Organisms characterised by cells with nuclei and mitochondria. Eukaryotes may be unicellular or multicellular.

Eusociality A form of social structure in which some group members permanently abstain from reproduction and invest their efforts instead in helping to rear the offspring of relatives. Common in ants, bees, and wasps.

Fitness A measure of evolutionary success. It can be thought of as the number of descendants an individual leaves or the number of copies of a gene that are successfully transmitted to future generations. It is a relative measure, used to compare a novel individual or a mutant gene with the norm or 'wild type'.

Gene A linear sequence of nucleic acids (DNA) that provides the information (code) required for (1) the production of proteins, (2) turning other genes on or off, and (3) making new copies of itself (replication).

Gene justice The Mendelian principle, enforced by meiosis, that alleles are represented in offspring in proportion to their frequencies in parents.

Genome The sum total of all an organism's genes and DNA.

Hamilton's rule A gene for helping a relative will spread if the cost to the helper (C) is less than the benefit to the recipient (B) multiplied by the degree of relatedness (R). Or, $C < BR$.

Horizontal transmission Movement of genes, other than between parent and offspring.

Inclusive fitness The sum of direct and indirect evolutionary advantage (fitness) gained through shared genes when an individual helps relatives.

Individual An independently reproducing unit. At **major transitions in evolution**, teams transform into new kinds of individuals.

Interdependence hypothesis The idea that members of a group have a share in each other's welfare (or fitness) and that this is the source of cooperation among non-kin.

Kin selection A form of natural selection, manifest through the benefit to relatives. Also see **inclusive fitness**.

Life A self-sustaining chemical system that is capable of Darwinian evolution.

Major transition in evolution The formation of a new kind of individual from the union of separate organisms or cells, which then reproduce as a unit.

Meiosis The process of cell division in which gametes (sperm and eggs) containing one set of chromosomes are produced from cells that contain two sets.

Mutualism A symbiosis between different species in which both benefit.

Mycobiont The fungal partner in a lichen symbiosis. See also **photobiont**.

Natural selection The process discovered by Charles Darwin which produces adaptation by the selective transmission of genetic variants from one generation to the next.

Organelle A discrete, membrane-bound structure inside a cell that carries out a specific function. Mitochondria and chloroplasts are organelles that originated as symbiotic bacteria.

Parasitism A symbiosis between different species in which one benefits at the expense of the other.

Photobiont The photosynthetic partner in a symbiosis such as a lichen. Either an alga or a cyanobacterium. Also see **mycobiont**.

Prokaryote A microbe belonging to the bacteria or the archaea.

Prosocial Behaviour that helps others, regardless of kinship.

Public goods Resources from which all may benefit, whether they contributed to their production or not.

Recombination A process by which genetic material is exchanged between homologous chromosomes, generating new combinations of genes. This occurs during **meiosis**.

Symbiosis A close association between dissimilar organisms that live together for a part or all of their lives. Symbioses may be parasitic or mutualistic, or exist along a continuum between the two.

Team A group of cooperators, bound together by the benefits that accrue from force of numbers and/or a division of labour.

Vertical transmission Movement of genes between parent and offspring.

Picture acknowledgements

Figure 1 - ProtoplasmaKid/Wikimedia Commons (CC BY-SA 4.0)

Figure 3 - The History Collection/Alamy Stock Photo

Figure 4 - Neil Bromhall/Shutterstock

Figure 5 - Mirrorpix/Mirrorpix via Getty Images

Figure 6 - O.S Fisher/Shutterstock

Figure 7 - © Prolineserver 2010, Wikipedia/Wikimedia Commons (CC BY-SA 3.0)

Figure 10 - Nancy R. Schiff/Getty Images

Figure 11 - Copyright © 2010, Oxford University Press

Figure 12 - Copyright © Cambridge University Press 2019

Figure 13 - Courtesy of FreeVintageillustrations.com (CC0 1.0)

Figure 14 - Image by Bernd Hildebrandt from Pixabay

Figure 15 - CNX OpenStax/Wikimedia Commons

Figure 17 - © 2022 Botanical Society of America

Figure 18 - © Copyright M J Richardson (CC BY-SA 2.0)

Figure 19 - © 2008 Hideko Urushiharar. Journal compilation © 2008 Japanese Society of Developmental Biologists

Figure 20 - From the National Oceanic and Atmospheric Administration Photo Library

Figure 21 - From the National Oceanic and Atmospheric Administration Photo Library

Figure 22 - © 2015 Terese Winslow LLC, U.S. Govt. has certain rights

Figure 23 - Copyright © 2016, American Association for the Advancement of Science

Figure 24 - Keystone/Getty Images

Figure 25 - Chiswick Chap/Wikimedia Commons (CC BY-SA 2.5)

Figure 26 - Tatiana Giraud, Jes S. Pedersen, and Laurent Keller's April 16 2002 article Evolution of supercolonies: The Argentine ants of southern Europe. Copyright © 2002, The National Academy of Sciences.

Index